STORM SPOTTING AND AMATEUR RADIO

THIRD EDITION

MICHAEL COREY, KI1U
VICTOR MORRIS, AH6WX
CONTRIBUTING EDITOR ROB MACEDO, KD1CY

Copyright © 2020 by

The American Radio Relay League, Inc.

Copyright secured under the Pan-American Convention

All rights reserved. No part of this work may be reproduced in any form except by written permission of the publisher. All rights of translation are reserved.

Printed in the USA

Quedan reservados todos los derechos

ISBN: 978-1-62595-141-0

Third Edition
First Printing

We strive to produce books without errors. Sometimes mistakes do occur, however. When we become aware of problems in our books (other than obvious typographical errors), we post corrections on the ARRL website. If you think you have found an error, please check **www.arrl.org/notes** for corrections. If you don't find a correction there, please let us know by sending e-mail to **pubsfdbk@arrl.org**.

Contents

Foreword

Acknowledgements

About ARRL

1 Introduction

2 Safety

3 Equipment and Resources

4 Training

5 Meteorology

6 Hurricanes

7 Storm Spotter Activation

Appendices

1 Weather Books for the Storm Spotter

2 Weather Websites

3 A Local SKYWARN Operations Manual

4 Memorandum of Understanding between the National Weather Service and the American Radio Relay League, Inc

5 False Statements Notice

6 Integrating Google Earth, NWS Data and APRS Using KML

7 WX4NHC: Amateur Radio station at the National Hurricane Center

8 Lightning Protection for the Amateur Radio Operator's Home

Index

Foreword

Amateur radio has played a part in responding to severe weather since the mid-1910s, providing communications when normal means go down and relaying emergency, priority and health and welfare traffic to assist various agencies at the local, state and federal level in this capacity. In addition to thias suppoart, the Amateur Radio Service and the American Relay Radio League, via a Memorandum of Understanding with the National Weather Service (NWS) SKYWARN program, have partnered to provide timely severe weather reporting into the NWS for the protection of life and property. The severe weather reportinag via SKYWARN covers all facets of weather such as hurricanes, tornadoes, severe thunderstorms, winter weather inacluding blizzards and nor'easters, floods, dust storms, coastal flooding and erosion, and even fire weather and volcanic ash. The severe weather reporting is then shared with other local and state government agencies, the media and the public at the discretion of NWS to create a situational awareness picture of what is really happening on the ground. The NWS SKYWARN program is open to all members of the public and most SKYWARN programs have both non-amateur radio and amateur radio SKYWARN spotters.

Having been out in the field covering the most dangerous severe weather events across the United States, the ground truth reports that I've received from amateur radio and non-amateur radio SKYWARN spotters allow me to know what is happening in and around the region where I am located. I can relay this information to the general public and tell them what actions they need to take to protect themselves and give them the hard facts of what is happening on the ground, and not just what radar and satellite tell me. The invaluable ground truth reports from tropical storms and hurricanes in the Gulf and Atlantic coast such as Hurricanes Harvey, Irma, Maria and Tropical Storm Nate in 2017 were invaluable in telling the story of what was really happening as these systems struck the area, and I could give viewers ground truth data on what was happening with these systems. Hurricane Lane affecting Hawaii with significant flooding and rainfall over 20 inches on the big island in August 2018 as well as Hurricanes Florence and Michael in September and early October 2018 in the Atlantic and Gulf coast regions were other examples of how spotter and amateur radio SKYWARN reports were invaluable to telling the story of what was really happening on the ground.

While I mention the reports with tropical systems, they aren't the only examples where reports from spotters and amateur radio were critical. When I have been out in the field reporting during severe thunderstorms and tornadoes across the country and coastal storm and nor'easter systems affecting the northeast, timely spotter reporting, and in some cases significant numbers of timely spotter reports, have provided me everything I need to tell the story of what is happening on the ground from the data and I did not just rely on technology. It better warns the public and help saves lives!

There are many resources available to allow amateur radio operators who are SKYWARN spotters learn more about severe weather, what to report, and become more effective. In this book, authors Mike Corey, KI1U, and Rob Macedo, KD1CY, look at all the various technology, resources, meteorological information, training and more to allow you find the right combination of tools to report safely and be safe during severe weather and how to properly respond during and after severe weather strikes.

I am grateful for all the SKYWARN spotters and amateur radio operators who are active and have provided reports into the NWS because they reach me and allow me to tell the story to the public so people know what is happening and can remain safe. If you are not yet active, we hope you will be inspired by this book to be an active spotter, be part of your local or regional SKYWARN group, and know that it means a lot not only to the NWS but to other government agencies, the media and the general public who want to know what is really happening at the ground and not just what radar, satellite imagery and models depict. It can save lives!

Jim Cantore
On-Air Meteorologist at The Weather Channel
Atlanta, Georgia
June 2020

Acknowledgements

A very heartfelt thanks to the University of Mississippi Amateur Radio Club, W5UMS; Oxford-Lafayette County SKYWARN; National Weather Service Boston-Norton; National Weather Service Memphis; and the WX1BOX SKYWARN team.

Very special thanks to Victor Morris, AH6WX, for his long time collaboration on this book, and to Rob Macedo, KD1CY, as a new collaborator.

SKYWARN® is a registered trademark of NOAA's National Weather Service

Amateur Radio Emergency Service® and ARES® are registered trademarks of the American Radio Relay League, Inc

This book is dedicated to my grandfather and mentor, the late Fred Selley, K9MXG.

About ARRL

We're the American Radio Relay League, Inc. — better known as ARRL. We're the largest membership association for the amateur radio hobby and service in the US. For over 100 years, we have been the primary source of information about amateur radio, offering a variety of benefits and services to our members, as well as the larger amateur radio community. We publish books on amateur radio, as well as four magazines covering a variety of radio communication interests. In addition, we provide technical advice and assistance to amateur radio enthusiasts, support several education programs, and sponsor a variety of operating events.

One of the primary benefits we offer to the ham radio community is in representing the interests of amateur radio operators before federal regulatory bodies advocating for meaningful access to the radio spectrum. ARRL also serves as the international secretariat of the International Amateur Radio Union, which performs a similar role internationally, advocating for amateur radio interests before the International Telecommunication Union and the World Administrative Radio Conferences.

Today, we proudly serve nearly 160,000 members, both in the US and internationally, through our national headquarters and flagship amateur radio station, W1AW, in Newington, Connecticut. Every year we welcome thousands of new licensees to our membership, and we hope you will join us. Let us be a part of your amateur radio journey. Visit www.arrl.org/join for more information.

225 Main Street
Newington, CT 06111-1400 USA
Tel: 860-594-0200
FAX: 860-594-0259
Email: membership@arrl.org

www.arrl.org

Introduction

It is a Tuesday night in early February. The National Weather Service has issued a statement that there is a probability of severe weather coming into the area late that night. They have been tracking a squall line several hundred miles to the west. The storm has been producing strong winds and severe thunderstorms. Reports indicate that it has also produced hail and isolated tornadoes. A hazardous weather outlook is issued and reads "SPOTTER ACTIVATION MAY BE NEEDED".

Long before the storm arrives, a local amateur radio SKYWARN® group is making preparations. Two stations volunteer to handle net control duties. Several members volunteer to go mobile if needed. Several more offer to relay weather reports from their home stations. Phone calls are made to the non-amateur radio spotters in the area to keep them current on what is happening. Everyone checks to make sure all is in order — radios working, batteries charged, flashlights handy, go-kits ready, and vehicles fuelled. The local ARRL Amateur Radio Emergency Service® (ARES®) Emergency Coordinator stays in touch with the National Weather Service and local emergency management, keeping up to date on the latest weather information. Now all that can be done is to wait.

February is not a typical severe weather month in this area. It is usually cool and rainy. But only a year before, almost to the day, a tornado outbreak struck the area killing 57 people and causing over a billion dollars in damages. The SKYWARN spotters were well aware of what was possible even in a normally slow severe weather month. As the spotters watch the line of storms approach they realize that this may be a very serious storm.

Before it arrives the net control stations activate the SKYWARN Net. One is operating from the local Emergency Operations Center, handling traffic coming in from the spotters. The other is relaying information to the SKYWARN Net one county to the east, keeping them abreast of the latest weather conditions.

Benjamin Rock, WX9TOR, storm spotting. [Benjamin Rock, WX9TOR, photo]

Mobile spotters are setting up to safely observe the storm as it approaches and are being repositioned as needed. Home-based spotters are calling in reports of conditions from their communities. Reports on the storm are being relayed to the National Weather Service office by radio, telephone, and Internet. Reports are coming in of wind gusts at 60+ MPH, ¼-inch diameter hail, street flooding, and a report of possible rotation in a cloud. Because of the incoming reports the National Weather Service can get a clearer picture of what is happening and can issue the appropriate watches and warnings.

The storm passes relatively quickly. In its wake come damage reports: trees down, power out, streets flooded, buildings damaged by wind. The spotter's job is not done. Calling in reports of damage is the next step. These after-the-event reports play a key part in understanding the weather. The local spotters, most also ARES members, continue to submit reports. The local emergency manager has also asked that they assist with search and rescue efforts in a flooded area and help provide communications for a local shelter that has been set up.

This scenario is not new for amateur radio operators who serve as volunteer storm spotters. Amateur radio has played a part in severe weather response for decades. But long before SKYWARN, radar, and the Internet, and even before amateur radio, there was a critical need to get real time, ground truth information on severe weather that could be used to warn of approaching storms and aid in forecasting. To understand how we got to where we are today, we should first look at severe weather in the United States, the history behind severe weather observation, and the role communications has played.

HISTORY

Weather in the United States is about extremes. One of the earliest weather observations tells us this. In the 1600s William Bradford, governor of the Plymouth Colony, noted this about winters in America: "sharp and violent, and subject to cruel and fierce storms, dangerous to travel to known places, and much more to search an unknown coast."[1] Throughout American history there are accounts of deadly floods, tornadoes, thunderstorms, hurricanes, and wildfires.

Regular observations of weather conditions began to be made starting in the 18th century. The founding fathers were keen observers of weather. Thomas Jefferson made regular observations from 1772 to 1778 and used a thermometer and barometer. He even noted on July 4, 1776, the temperature in Philadelphia was 76°. George Washington kept daily weather observations in a diary until the day before his death. And Benjamin Franklin's weather experiments are part of popular American folklore.

In the 19th century a major advancement in communications, the telegraph, made it possible to relay weather observations to other stations. This was the starting point of weather forecasting. Advance notice of approaching weather had always been a need, but it was impossible without the ability to communicate real time observations. It was not long after the invention of the telegraph that organized observation of weather commenced.

The Birth of a National Weather Service

In 1848 Professor Joseph Henry, Secretary of the Smithsonian Institution, proposed a program to observe and report severe weather. The program, The Smithsonian Institution Meteorological Project, would rely on a network of volunteers across the country. By the end of the first year the project had more than 150 volunteers and by 1860 there were more than 500. The project relied on the telegraph system to relay information from volunteer observers to the Smithsonian. In 1862 the Smithsonian published a pamphlet containing information about tornadoes and thunderstorms. The public was urged to take note of weather occurrences and forward the information on to the Smithsonian. As time went on other weather observation systems around the country, such as that organized by Professor Cleveland Abbe in Cincinnati,

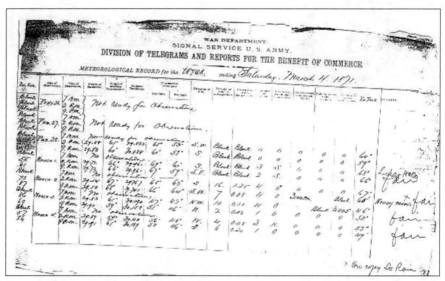

The first week of official observations taken in Memphis, Tennessee. [courtesy National Weather Service Memphis WFO]

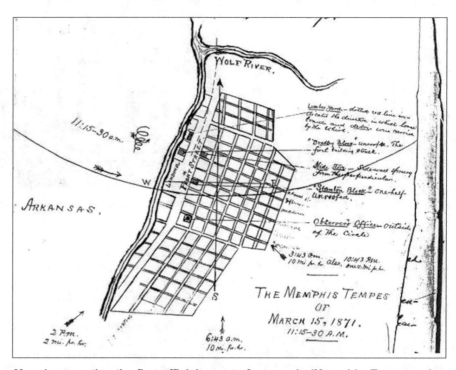

Map documenting the first official report of a tornado (Memphis, Tennessee). [courtesy National Weather Service Memphis WFO]

would be absorbed into the Smithsonian's network.

During this time Professor Increase A. Lapham, a scientist, author, and student of meteorology, urged that the federal government establish a weather service that could coordinate weather observations and issue forecasts. With the support of Colonel Albert Myer of the US Army Signal Corps, he succeeded in lobbying Congress to pass legislation establishing a national weather service. In 1870 President Ulysses S. Grant signed a bill that moved the weather service from the Smithsonian to the Department of War. The new department was assigned to the Signal Corps and given the name The Division of Telegrams and Reports for the

Benefit of Commerce. The use of volunteer observers was replaced with Observing-Sergeants who would telegraph weather reports to Washington DC. Initially there were 24 observing stations. By 1878 there were 284.

Along with the shift from the Smithsonian to the Signal Corps came the first organized weather training. There were very few professional meteorologists in the Signal Corps. The observation stations were made up of military personnel who had little or no background in weather spotting. At Fort Whipple in Virginia a course was developed within the curriculum of telegraph and signaling that covered meteorology and meteorological observation. Eventually a similar course was developed for officers. The training course lasted until 1886 when it was ended by the Secretary of War.

Due to internal turmoil, President Benjamin Harrison moved the weather service to the Department of Agriculture in 1890. The change created a new civilian weather bureau. The military personnel who had staffed the weather service under the Signal Corps were discharged and given the opportunity to go to the new department. With the transfer also came a name change: from 1891 to 1967 it would be known as the Weather Bureau.

During this time, technological advancements greatly influenced how weather observations were reported and the ability to forecast weather. In 1902 the Marconi Company began sending forecasts from the Weather Bureau to ships in the Cunard Line. And in 1905 the first weather observation from a ship at sea was sent by wireless. Wireless was a huge step forward over sending reports by telegraph. Telegraph required observation stations to be located along telegraph lines, while with wireless, observation stations could be placed anywhere. As wireless advanced, so too did the ability to receive real time weather observations. In 1921 the University of Wisconsin made the first transmission of weather forecasts by voice. By 1928 teletype had replaced voice for sending weather information.

Also during the early years of the Weather Bureau the use of volunteer observers re-emerged. In 1890 the Cooperative Observer Program was established and is still in use today. The program has two goals: "to provide observational meteorological data, usually consisting of daily maximum and minimum temperatures, snowfall, and 24-hour

The *Severe Storm Reporting Handbook* (1956) and *It Looks Like a Tornado* (1959). [courtesy National Weather Service]

precipitation totals, required to define the climate of the United States and to help measure long-term climate changes, and to provide observational meteorological data in near real-time to support forecast, warning and other public service programs of the National Weather Service."[2] In 1933 a science advisory group indicated to the President Franklin Roosevelt that the use of a volunteer cooperative observer network provided an extraordinary service. Today the program has more than 8700 volunteers.

In 1940 the Weather Bureau moved to another department within the federal government. President Roosevelt moved it to the Department of Commerce, where it remains today under the name National Weather Service, because of the role that it played in aviation and therefore the nation's commerce. During the years of World War II, the Weather Bureau, along with the military, set up a network of volunteer storm spotters. The initial mission of the spotters was to monitor lightning near ordnance facilities. The value of real time, first-hand information was apparent and by 1945 there were more than 200 observer networks in the United States.

The use of weather observers continued after the war. A series of tornadoes between 1947 and 1951 in Texas showed that having a local storm spotting network made it possible to get warnings out faster and save lives. Spotter networks continued to develop throughout the late 1940s and early 1950s. The use of spotting networks had a noticeable impact on the number of tornadoes being reported.[3] In May 1955 a tornado struck Udall, Kansas, killing 80 people and injuring 273. This event prompted the Weather Bureau to recruit spotters and develop training for them. On March 8, 1959, the first official storm spotter training was conducted in Wellington, Kansas. A group of 225 spotters received training. Along with the training the Weather Bureau issued a handbook for spotters, the *Severe Storm Reporting Handbook* (1956). This was followed up in 1959 with the publication *It Looks like a Tornado*.

In 1965 another weather event impacted storm spotting networks around the country. On April 11, 1965 — Palm Sunday — 47 tornadoes struck the states of Indiana, Illinois, Wisconsin, Ohio, and Michigan. This outbreak was the second biggest tornado outbreak on record at that time and the deadliest tornado outbreak in Indiana with 137 killed. Over a period of 12 hours, 271 people were killed and more than 3500 injured across the five states.

Following the Palm Sunday tornado outbreak, the Weather Bureau realized that there were serious shortcomings in early warning, communications, and storm spotting. From this came the National Disaster Warning System or NADWARN. A phase of NADWARN was the SKYWARN program. The SKYWARN program would be coordinated by the Weather Bureau (to become the National Weather Service in 1967) and train local volunteer storm spotters. Training classes were developed and the use of publications about severe weather was continued. SKYWARN started officially in the early 1970s. Since its beginning it has expanded to cover severe weather beyond tornadoes. Today there are between 350,000 and 400,000 SKYWARN trained storm spotters in the United States. The concept of SKYWARN has also made it to other countries, with groups such as TORRO in the United Kingdom, SKYWARN Europe, and CANWARN in Canada.

Amateur Radio and Storm Spotting

Amateur radio has played a part in responding to severe weather long before the development of SKYWARN, even before the development of organizations such as the ARRL's Amateur Radio Emergency Service (ARES). Starting in the 20th century we see how amateur radio operators were involved when severe weather struck. During these early years the role was primarily in response to severe weather instead of ahead of it. Amateurs provided communications when normal means went down, relayed health and welfare traffic, and assisted various local, state, and federal agencies during recovery. If we look at some of the stories from these early years we can see many similarities to what we do today.

The first instance of amateur radio's value to the public involved a weather event. In 1913, radio amateurs in Michigan and Ohio passed disaster messages when other means of communications were down in the aftermath of severe storms and flooding. A Department of Commerce bulletin followed, proposing a dedicated communications network of radio amateurs to serve during disasters. Five special licenses were reportedly issued. A magazine article noted that amateurs — who were once considered nuisances — were now considered to be essential auxiliary assets of the national public welfare.

In 1923 the Arkansas River, which crosses northeast Oklahoma, flooded. Normal lines of communication between Tulsa and the nearby town of Sand Springs were lost. Amateur radio operators Edward Austin, 5GA, Raymond McKinney, 5SG, and John Lewis, 5WX, of Tulsa, along with Halton Friend, 5XBF, and Howard Siegfried, 5GJ, of Sand Springs used amateur radio to maintain communication. The operators handled messages between the towns and relayed information from reporters to their newspapers. The operators continued this for three days, until normal communication lines were restored.[4]

On January 8, 1937, an ice storm struck the central states of Missouri, Illinois, Arkansas, and Oklahoma. Across the region communication lines were down. Amateur radio operators in the affected areas assisted by providing communications for almost a week before normal communication lines were restored. Amateurs handled traffic for utilities such as electric and gas companies, telephone companies such as Southwestern Bell, the Associated Press, Western Union, and railroads. One station, W9PYF, logged 103 hours of operation, handing 183 messages.[5]

In September 1941 a hurricane struck the Texas Gulf coast. The hurricane, a category 3, known as Hurricane 2 (naming of hurricanes was not standardized until 1951) struck between Freeport and Port O'Conner on September 23. The storm caused a great deal of damage and took out communication lines in the areas around Galveston and Houston. An amateur radio operator, James "Salty" Johnson, W5LS, of Houston, Texas volunteered to assist with communications. Before the storm made landfall he took a truck containing a 300 W transmitter, generator, and receiver and went to the area where it was thought the center of the storm would strike. He waited the storm out in the city jail at Palacios, Texas. Salty was not alone; other amateur radio operators along the coast set up and prepared for the storm and to provide communications after it made landfall. Amateurs from around the state and around the country were ready to assist with disaster communications.[6] The death toll was only four. This was attributed to early warning and preparedness.

On the morning of September 29, 1938, a storm system moved into the Carolinas from the Gulf of Mexico. The storm produced five tornadoes that struck South Carolina. Two of the tornadoes were rated as F2 and three were rated as F1, resulting in 32 killed and around 100 injured. Assisting in the response to the devastation in Charleston, South Carolina were local amateur radio operators. As soon as power was restored to some areas amateurs went on the air and began handling traffic. Although telephone lines were not completely down they were overloaded by people calling to reach family and friends. Amateurs also assisted local law enforcement with communications for the police force and rescue efforts.[7]

Prior to World War II, most of amateur radio's involvement with severe weather was in response to incidents. The reason for this was the limited ability to forecast weather. Most Weather Bureau offices were still relying on little more than a thermometer, anemometer and wind vane, and barometer to detect changes in the weather. Innovations such as radar and weather satellites would not come until after the war.

The first effort at organizing any form of amateur radio network to do storm spotting came during World War II. The War Emergency Radio Service (WERS) was started by the Federal Communications Commission in June of 1942. The regular amateur radio service was suspended during the war, but there was still a need for communications that amateur radio operators could fill. WERS was made up of licensed radio amateurs who operated WERS stations that were licensed to a community. The purpose was to assist with air raid warning and eventually expanded to include natural disasters and severe weather. Severe weather spotting was a secondary function that developed during the war. The original idea was to report on severe weather near military ordnance facilities. The weather observing component of WERS

W.E.R.S. EXPANDED

On April 10th some important changes were made by FCC in the regulations governing WERS, Part 15 of the FCC Rules, acting upon a request filed by ARRL about a year ago. As the regulations have read, WERS stations must be licensed primarily in connection with national security and defense and, although the regulations were changed some time ago to permit *such* stations to operate during natural-disaster emergencies, no provision has hitherto existed for the organization of WERS primarily on behalf of the relief of disasters. The regulations are now expanded to permit the CD type of WERS stations to operate in an emergency jeopardizing public safety, and new networks may be established and existing licenses renewed even though the Citizens' Defense Corps or equivalent civilian defense organizations are no longer active in the areas involved. WERS is the only agency through which amateur services may be utilized in the traditional duty of the amateur to cope with communication difficulties flowing from public emergencies, so this is a very significant expansion, as we comment upon in this month's editorial.

Article from the June 1945 *QST* on the War Emergency Radio Service.

was formalized in 1945.

Following World War II, most amateur radio activity related to severe weather was still in providing assistance after the storm struck. Amateurs continued to set up emergency nets to relay traffic and assist with rescue efforts. Throughout the post-war years a combination of events changed the way amateur radio was involved in severe weather. Forecasting severe weather was advanced by the use of radar and weather satellites. The first tornado warning was issued, setting a precedent for future weather warnings. Amateur radio technology was advancing to make it easier for stations to operate mobile and portable. Many of the storm spotter networks that started in World War II were still active and new ones were being formed. Storm spotter training was beginning to take off. When SKYWARN finally came into being, it was a perfect match with amateur radio.

STORM SPOTTING TODAY

Storm spotting has continued to evolve and advance since the start of the SKYWARN program. Originally, the primary concern for storm spotters was tornadoes and thunderstorms. These severe weather events still make up the majority of storm spotter activations. Today storm spotters are also activated for a wide range of severe weather events such as hurricanes, winter weather, floods, dust storms, and nor'easters. Spotters are also activated for geologic issues such as coastal erosion and volcanic ash fall.

Over the last 30 years there have also been monumental advancements in technology that have benefited storm spotting. Within the amateur radio community the growth in repeaters on the 2 meter and 70 centimeter bands have made running SKYWARN nets possible almost anywhere in the country. And new modes such as EchoLink, digital voice, and APRS have helped to fill in communication gaps and generally make storm spotter networks more robust.

Outside of amateur radio there have been many advancements that have helped volunteer storm spotters become more effective. Smartphone technology today gives every user the capability to stay aware of severe weather and weather forecasts. Computers have become an essential tool for the storm spotter. Combined with modern weather software packages they give the storm spotter or net control operator access to data not available 30 years ago. Dissemination of weather information has increased exponentially. Today weather information can arrive via apps, television, radio, satellite, NOAA Weather Radio, text message, email, and websites. Text messaging and social media programs make it possible to network spotters, net control stations, emergency management, meteorologists, and news media seamlessly.

With the advances in technology also come advances in what we have learned from severe weather. Regardless of your role in the severe weather warning process — spotter, public safety official, NWS, news media — with each severe weather event we learn from it and learn how to respond better in the future.

The ability to forecast severe weather has also evolved greatly over the last 30 years. Advances in radar, communications, networking, and research have made forecasting more accurate. For the amateur radio operator who is a storm spotter the amount of available technology, resources, meteorological information, training, and the impact from severe weather can all seem daunting. In this book we will look at these things and see how amateur radio operators can apply the right combination of tools to be safe during severe weather, respond to it, and assist in its aftermath.

SKYWARN Recognition Day

SKYWARN™ Recognition Day was developed in 1999 by the National Weather Service and the American Radio Relay League. It celebrates the contributions that SKYWARN™ volunteers make to the NWS mission — the protection of life and property. Amateur radio operators comprise a large percentage of the SKYWARN™ volunteers across the country. The Amateur radio operators also provide vital communication between the NWS and emergency management if normal communications become inoperative. During the SKYWARN™ Special Event, operators will visit NWS offices and contact other radio operators across the world.

SKYWARN Recognition Day is not a contest. The object is for stations to exchange some basic information with as many NWS stations as possible on 80 meters through 70 centimeters (excluding 1.25 meters and 30/17/12 meters). Repeater and EchoLink/IRLP contacts are permitted. Stations exchange call signs, signal reports, locations, and a one or two-word description of the weather at their respective locations.

Each year nearly 70 amateur radio stations go on the air from National Weather Service offices around the United States. Activity is on all the traditional modes; voice, data, and CW. Details of SKYWARN recognition day are published each year through ARRL and NWS.

SKYWARN Recognition Day at National Weather Service Taunton, Massachusetts, station WX1BOX (currently located at the NWS Boston/Norton Forecast Office). [Mike Corey, KI1U, photo]

OUR PURPOSE, AUDIENCE, AND ASSUMPTIONS

Before we get into the heart of the material we need to cover the purpose and intent of this book and for whom it is written. We also have to go over certain assumptions that will be made throughout this book. And finally we need to make some clear distinctions about storm spotters.

The purpose of this book is *not* to compete with the NWS's official SKYWARN spotter training. Storm spotters, whether they are amateur radio operators, public safety officials, or just concerned citizens, must be trained by meteorologists. We must remember that the purpose of a storm spotter is to relay ground truth information to those who issue our nation's warnings and forecasts, the National Weather Service. This is why the NWS developed SKYWARN training and why NWS meteorologists conduct SKYWARN training. Granted, there are other training programs out there designed for storm spotters, but all will generally agree that the basis for training comes with the basic SKYWARN course.

Many who go through the National Weather Service (NWS) SKYWARN course are amateur radio operators, this is not by accident. Amateur radio and storm spotting are a great match. amateur radio operators bring to storm spotting great resources such as an established, resilient and robust communications system that can function in an emergency or disaster; a pool of volunteers willing to be trained; a history of public service; and technologies that no other group has — all paid for by the individual amateur.

The purpose of this book is to be a resource for the amateur radio operator who volunteers as a trained storm spotter. It presents information on resources, training, equipment, safety, storm spotter activation procedures, reportable weather criteria, developing a local storm spotter manual, and the experiences of storm spotters from around the country. It also provides some meteorological information about severe weather such as hurricanes, tornadoes, hail, floods, damaging wind, and winter weather.

The primary audience in mind is the amateur radio community. It is not just for experienced storm spotters, but also for those new to this aspect of the hobby or interested in getting into storm spotting. Non-amateurs may also find some material useful. While they are not the intended audience, if the material presented helps

20 Questions for a Storm Spotter
Answered by Rob Macedo, KD1CY, New Bedford, Massachusetts

Q1. How long have you been a storm spotter?
A. I have been interested in weather spotting and running weather/SKYWARN nets since 1990. I received SKYWARN training from the NWS Taunton (now Boston/Norton) Forecast Office in 1995 and have been refreshed in the classes many times over the years. I am now an instructor of the SKYWARN training classes for the NWS Boston/Norton Forecast Office and have been an instructor since 2008.

Q2. What types of severe weather are you activated for?
A. We activate for severe thunderstorm/tornadic events, hurricanes/tropical storms and tropical systems, winter storms and blizzards, high wind events — virtually every form of severe weather.

Q3. What type of training have you had?
A. We have a comprehensive 2 – 2.5 hour SKYWARN Training class. We do not have a basic and advanced SKYWARN class like a number of other National Weather Service Forecast Offices.

Q4. Approximately how many storm spotters are in your area?
A. There are more than 8000 SKYWARN Spotters in the NWS Boston/Norton coverage area. In the local area where I live, we have several hundred spotters.

Q5. How does information get from the spotter to the NWS?
A. We maintain a SKYWARN Announcement email list that has close to 3000 SKYWARN spotters and push information on SKYWARN Activations, Storm updates, SKYWARN and other training via that email list. We also have social media accounts for Twitter and Facebook under WX1BOX. The NWS office also has social media outlets to push information and will share our reports, pictures and videos when posted.

Q6. Does the local storm spotter group participate in any drills or exercises?
A. The NWS Boston/Norton office will participate in various drills and exercises when they are set up in the region.

Q7. How does your local SKYWARN group include non-ham spotters?
A. Our email list, and Facebook and Twitter social media accounts are ways we work with and keep in touch with spotters who are not ham radio operators. Some of these spotters get their ham radio license.

Q8. How often do you have local SKYWARN training?
A. The SKYWARN training is given in the Spring through Fall of each year across the NWS Boston/Norton coverage area. Local locations are hit once every few years in some areas and once every year or two in other locations that are centrally located and drawn more people than other areas.

Q9. Describe the relationship between your SKYWARN group and local emergency management.
A. Emergency Management will often use SKYWARN reports for situational awareness and disaster intelligence gathering. Emergency management and other public safety agencies will also share damage and weather information with us as well.

Q10. What amateur radio modes and frequencies do you use for storm spotting?
A. We use largely 2 meters/440 MHz as well as EchoLink and IRLP. HF can be used in situations such as hurricanes and tropical systems. We are reviewing the use of DMR, D-STAR and NBEMS as other modes of operation, with liaisons to those modes being explored as it's difficult to monitor all the various modes from the local National Weather Service Office with just two or three operators.

Q11. As a volunteer storm spotter what are your primary safety concerns?
A. Our primary safety concerns are any damaging type weather, ranging from severe thunderstorms and tornadoes, to flooding of both freshwater and coastal areas, to high wind and winter storm/blizzard events. There can also be unique situations such as the winter of 2014 – 2015 where the heavy snow that fell from multiple major winter storms and blizzards caused many structural roof and building collapses in our area. SKYWARN was responsible for many of those reports, especially when they occurred during the winter storms and blizzards.

them as a storm spotter or perhaps generates interest in amateur radio, then that would be welcome.

There are certain assumptions that we must make throughout this book: definition and purpose of a storm spotter, storm spotter affiliation, function of storm spotter reports, the unique role that amateur radio plays, and fundamental training of all storm spotters. There is no single, set definition of what a storm spotter is. We can define a storm spotter, though, in these terms: a storm spotter is a volunteer who visually monitors specific weather conditions and their progression and relays that information to a local weather office.

In the United States these volunteers are trained by the National Weather Service through the SKYWARN weather spotter program. The NWS defines the responsibility of the storm spotter as "identify and describe severe local storms."[8] The purpose of having this pool of trained volunteers is to provide ground truth reports which along with other partners, such as media, storm chasers, public safety officials, and emergency management, make up a reporting network. Information from this network is used, along with information from radar and other sources, in issuing severe weather watches, warnings, and advisories.

Spotters trained and certified by the NWS are considered volunteers to the Federal Government. A non-paid volunteer observer may be considered an "employee" under the Federal Employees Compensation Act. Final determination rests with the Department of Labor's Office of Workers Compensation Programs. Any spotter injured while providing observational duties should notify their local NWS office. The local NWS office and their regional office should work with the Office of Workers Compensation Programs for resolution.[9]

It is absolutely critical that we understand an important concept about storm spotting. Storm spotters are *not* storm chasers, and likewise, storm chasers are not storm spotters (see the sidebars). These are two different activities that are often, and mistakenly, considered to be the same thing. There are very critical differences between these two activities.

First is the difference in what weather conditions each one is looking for. Storm spotters are observing severe weather conditions in their area. The weather conditions that are reportable are set by the local NWS office. Storm chasers are actively seeking and going after severe weather that may be occurring far away from where they live.

Training is another difference. Storm spotters are almost always trained by the NWS SKYWARN program. They may also receive additional training in severe weather through advanced SKYWARN or

Q12. In your area, are the evacuation centers, Red Cross, EOCs, and so on equipped with amateur radio?
A. It is variable across the area. Some Red Cross shelters/evacuation centers and EOCs have amateur radio equipment but some don't. It depends on the local community and how active the Red Cross, Emergency Management, and similar agencies are when it comes to amateur radio.

Q13. Describe how your local SKYWARN net is conducted.
A. It varies. In some cases, a net control and formal net is conducted in the case of a Severe Thunderstorm/Tornado warning. In winter storms, net call-ups can occur every hour or few hours depending on the rate of snowfall and other weather-related issues associated with the storm system. In other cases, WX1BOX, the amateur radio station at NWS Boston/Norton, will periodically poll for reports on area repeaters if there is no volunteer net control or formal net called. During what we call self-activation events, we will often be in a monitoring mode from our home location looking to gather reports from the various areas. These events we self-activate for may be localized or confined to be predominately at one weather hazard event such as accumulating snow or strong to damaging winds as a couple of examples.

Q14. Does your local SKYWARN group conduct or participate in any training that is not SKYWARN?
A. Sometimes, we have Emergency Communications Workshop training that is offered by our section and SKYWARN spotters and groups will participate.

Q15. What other amateur radio activities are you involved in?
A. I am also involved with public service events such as the Boston Marathon and was the ARRL Eastern Massachusetts ARES Section Emergency Coordinator for nine years (2005 – 2014), ARRL Assistant Section Emergency Coordinator from 2014 – 2018 and I am now back as the ARRL Section Emergency Coordinator since 2018.

Q16. What got you interested in storm spotting?
A. When I was a kid, I used to listen to KA1JJM — Ray Weber and the Western Massachussetts SKYWARN Net on the 146.910 Mount Greylock repeater during severe weather events. That inspired me to get my amateur radio license and take an interest in SKYWARN storm spotting. My Uncle, who was a ham (Louis Macedo, W1WAY) also supported nets in Massachusetts during the hurricanes of the mid-1950s to early 1960s that affected New England.

Q17. Do you have a go-kit? What's in it?
A. I have six radio go-kits. Two are VHF/UHF and digital packet/NBEMS (Narrow Band Emergency Messaging System) capable, one is HF/VHF/UHF with NBEMS capability and three are DMR Go Kits, 1 VHF, 1 UHF, and 1 Dual Band VHF/UHF. I also have other go-kit materials so that I can deploy and have key items to deploy to a shelter, EOC or other location as needed.

Q18. What, in your opinion, is the most valuable tool a storm spotter can take with them when spotting?
A. Their eyes and being observant, remaining calm when reporting conditions, and assuring the precision of such reports as they relay them into the National Weather Service.

Q19. How can amateur radio storm spotters improve?
A. Be an active spotter. When conditions are ripe for severe weather, report into an amateur radio net or report by other means. Never assume the National Weather Service and other agencies are aware of the damage that occurred if you observe it. It is much better to report in and ensure that the National Weather Service and other key agencies have the information.

Q20. When you are activated how do you participate in SKYWARN (as mobile spotter, net control, relay, and so on)?
A. Most often, I'm located at the National Weather Service in Norton, Massachusetts. gathering reports from established nets and running nets where required. Occasionally, I'll operate from home performing the same function and sometimes I've also acted as a mobile spotter reporting in conditions and taking pictures of criteria conditions for the NWS.

through local emergency management or other sources. Storm chasers, many of whom are meteorologists, generally have a broader, more in-depth knowledge of severe weather. This is not to say that spotters are less knowledgeable, but that the difference in what each one does requires different levels of meteorological knowledge.

Storm spotters are coordinated through the local NWS office. When spotter activation is necessary, the local office will let spotters know through hazardous weather outlooks, email, text messaging, or via local emergency management. At the community level spotters are often coordinated through amateur radio clubs, emergency coordinators, SKYWARN coordinators, or emergency management. Storm chasers most times operate independent of any organization although some are coordinated through a university or research center. While in the field, chasers may coordinate and share information with other chasers.

Equipment is also a major difference. Storm spotters generally use basic equipment that helps them monitor local weather conditions. This may include amateur radio, cell phone, camera, and GPS. Storm chasers often use professional grade equipment to monitor a wide range of weather conditions. Storm spotters operate

20 Questions for a Storm Chaser
Answered by Benjamin Rock, WX9TOR, Woodstock, Illinois

Q1. How long have you been a storm chaser?
A. About 16 years.

Q2. What is your educational / training background?
A. Most of it is self taught, but I have taken many classes in earth science, meteorology, math, chemistry and geography.

Q3. How would you define the difference between storm chasers and storm spotters?
A. Spotters wait for the storm to come to them and then report their findings to the NWS by some means. Chasers will track a storm down, tape it or get photos of it. Some do it for research and others for the adrenaline rush or personal gain.

Q4. Describe the relationship between storm chasers and storm spotters.
A. Most chasers started out as spotters and evolved into chasers for different reasons.

Q5. While chasing, do you monitor Amateur Radio communications?
A. Always.

Q6. What equipment do you take chasing?
A. Laptop with Internet, water and snacks, cell phone, go-kit, camera, batteries, ham and CB radios, NOAA radio, pens, paper, and basic office supplies. We used to pull over and use a pay phone to call in reports and had paper forecast charts that were printed out the night before to use. It sure has come a long way.

Q7. What got you interested in chasing?
A. The science of how a tornado is formed. It's pure fascination.

Q8. When chasing, do you coordinate with local public safety and emergency management officials?
A. Yes.

Q9. How do you relay information to NWS?
A. Via ham radio or phone.

Q10. As a chaser what are your primary safety concerns?
A. The other people out there when we are chasing. They tend to stop suddenly and drive toward what they are staring at.

Q11. What types of severe weather do you chase?
A. Thunderstorms and tornadoes.

Q12. In your opinion, do you think storm chasers can help in training storm spotters?
A. Yes, we have real-time experience with these storms that most spotters may never see.

Q13. How many are in a chase team and what are their roles?
A. The number varies. My team has three. Driver, navigator / radar man, and cameraman. Most chasers, however, meet up with many others in the field so the number may be as high as eight or more in a vehicle doing different stuff.

Q14. What geographic areas do you storm chase in?
A. I've chased from coast to coast and border to border.

Q15. When not chasing what do you do? (Meteorologist, academic, or other occupation?)
A. I have a full time job at the local newspaper.

Q16. What kind of relationship is there between storm chasers and media?
A. That's tough. We all have certain media outlets we use to sell video or pictures for the local news. We very rarely go outside that circle.

Q17. How do you determine when to go on a chase?
A. Well, if the models show a decent area for storm development, the decision is made based on models and gut instinct. Usually you know a couple of days in advance.

Q18. Is there specialized training for storm chasers?
A. Not really. Most are self taught.

Q19. How realistic are TV shows about storm chasing?
A. Like the show Storm Chasers on Discovery Channel? I would say it's a fifty-fifty mix. They edit out the days you bust because it doesn't make good watching. There are days on end where you drive and drive hundreds of miles to get to an area that looked great for development a couple of hours earlier only to sit for hours and do nothing and have the storm die out before you see anything. They never show that . . .

Q20. In what ways can the field of storm chasing improve?
A. That's hard to answer. There are so many factors taken into account for chasing on any given day that to narrow it down is nearly impossible. It's not a perfect science. I guess the image itself of a storm chaser could be improved. There are so many people who have seen that famous movie and watched all those shows and then saw that guy get so close to the tornadoes that he could touch it. Because of this, they are perceived as reckless, dangerous, law-breaking people. Others who don't have experience tag along behind our groups and think they can do it too. They don't realize how dangerous it truly is until it's too late and someone doesn't come home.

within in local area. They are trained to report on conditions unique to where they live. Storm chasers may travel hundreds or thousands of miles to chase severe weather.

The information gathered by spotters and chasers can be used for a variety of purposes. A spotter's primary responsibility, as stated earlier, is to relay important weather information back to the local NWS office. This may be done directly or via a net control station, relay station, or emergency management. Storm chasers may also report weather information to the local NWS office, but often the data they are gathering may be used for research purposes, commercial purposes, or individual interests.

There is one critical difference that every storm spotter should keep in mind. While the NWS does support storm spotters through the SKYWARN program, it does not support storm chasing. It is absolutely critical that every storm spotter know this and know the boundaries between spotting and chasing. This book is intended for storm spotters, not chasers.

Amateur radio and SKYWARN have enjoyed a solid relationship since the inception of SKYWARN in the late 1960s. This relationship has benefited the amateur radio community by providing training, a regular source for public service, realization for the need to advance communication capabilities, and a solid relationship, not only between the ARRL and the NWS, but all the way down to the local amateur radio club and the local NWS office. For the NWS this relationship has provided a large pool of volunteers who come with a communications network designed for emergency use, a vast increase in the number of storm reports, improved communication during severe weather, and another link to the local community. Simply put, without spotters the NWS would not be able to fulfill its mission of protecting life and property.

Now let's get into the heart of the material. The first issue we need to address, and will continue to address throughout this book, is safety. Safety is *always* our #1 priority.

REFERENCES

[1] D. Ludlum, "Early American Tornadoes, 1586 – 1870," American Meteorological Society, 1970.
[2] National Weather Service, "Cooperative Observer Program," **www.weather.gov/coop/**.
[3] C. Doswell, A. Moller, H. Brooks, "Storm Spotting and Public Awareness Since the First Tornado Forecasts of 1948," *Weather and Forecasting*, 14, 544 – 557, Aug 1999.
[4] H. Mason, "Amateur Radio Furnishes Communication During Flood," *QST*, Aug 1923, p 54.
[5] "Amateurs Provide Communication During Ice Storms," Operating News, *QST*, Mar 1937 pp 53-55.
[6] C.B. DeSoto, "Texas Hurricane Finds Hams Ready," *QST*, Nov 1941, pp 39, 94, 96.
[7] "So Carolina Tornado Emergency," *QST*, Jan 1939, pp 94, 96.
[8] National Weather Service, "NWS SKYWARN Storm Spotter Program," **www.weather.gov/skywarn/**.
[9] Federal Employee's Compensation Act, 5 USC Chapter 81

Safety

On May 10, 2008, a series of supercells developed over northeast Oklahoma, southeast Kansas, and southwest Missouri. The storms produced hail up to softball size, lightning, damaging winds, and 11 tornadoes. The storm system was part of a larger tornado outbreak that spawned 147 tornadoes over 18 states between May 7 and May 15. Altogether the outbreak resulted in 25 deaths, 22 of the fatalities occurred in Picher, Oklahoma, and Newton, Jasper, and Barry Counties in Missouri.

One of the fatalities was a storm spotter. Tyler Casey, a volunteer firefighter from Seneca, Missouri, was killed when his vehicle was thrown by a tornado. Tyler was spotting near the intersection of Highway 43 and Iris Road, not far from where an EF-4 tornado touched down.

The purpose of mentioning Tyler's untimely and unfortunate death while serving his community is not to open it up for critique or review. It is, instead, to drive home the point that severe weather is dangerous and can injure and kill. Training, command and control, and communication are all critical for spotters to maintain situational awareness. When we serve as storm spotters, whether at home, from a net control location, or while mobile, we take certain risks. Our first duty, and a duty above any other (including submitting reports), is safety. We must keep ourselves safe and we must not do anything that would jeopardize anyone else's safety. Safety is a topic we can never go over too much. It should always be on our mind when serving as a volunteer storm spotter.

We must keep in mind that storm spotters do not have to operate mobile. Being mobile does give us the advantage of observing incoming weather from a better vantage point and the ability to leave the area if it becomes too dangerous. But overall, storm spotting while mobile is far riskier than spotting from a fixed location. There are times that storm spotting while mobile is too dangerous to do no matter how experienced or well trained the spotter may be. Hurricanes, low visibility conditions such as fog or dust storms, blizzards, and spotting at night are all

A windmill standing alone against a supercell thunderstorm in Leedey, Oklahoma, on April 17, 2002. [Stephen Corfidi, NOAA/NWS/SPC/OB, photo]

times when a spotter should not venture out but report from a secure location.

Just because we are in a house or some other structure does not guarantee safety. Generally we are safer inside a building than in a car, but no building is immune to the effects of weather. Homes, fire stations, hospitals, police stations, and even emergency operations centers have all been damaged or destroyed by severe weather. In 2005 the Knight Township Volunteer Fire Department in Vanderburgh County, Indiana, was heavily damaged by a tornado.[1]

In this chapter we will look at some guidelines that can help keep a storm spotter safe. The National Weather Service, working with the American Red Cross, has developed spotter safety guidelines that are a key part of the SKYWARN training program (see the sidebar later in this chapter). Beyond that there is safety advice that should be heeded from a public safety standpoint. And there are lessons learned from those who have served as spotters and even from the storm chasing community. We will look at all of these and how they apply to storm spotting. We will then look at safety recommendations that are specific to different types of severe weather. And finally we will look at items you can keep in your car or in your home that will help keep you safer.

MOBILE SAFETY

First let's look at safety issues for the mobile storm spotter. A mobile storm spotter, while able to gain the best vantage point for observing weather conditions, is also at the greatest risk if something goes wrong. A vehicle provides poor protection against hail, wind, and tornadoes. And getting stuck in a dangerous situation is a possibility. Storm spotters can help minimize some risks by following certain safety guidelines, taking some preparatory measures, and making sure their vehicle is properly equipped.

Buffer Zone

When spotting while mobile you should always keep a buffer zone between you and the storm. By keeping a safe distance you have more room to maneuver if the storm changes direction. Also it allows you to keep an open escape route if you need to leave the area. In the accompanying photo the spotter sets up in the rear right flank of the storm. Since the storm is moving from the southwest to northeast the spotter is in a vantage point to observe the storm without being directly in its path. They are still able to observe reportable conditions. This location also gives several escape routes if the spotter needs to leave the area.

Understanding the buffer zone is critical to know what else may be coming. There may be another dangerous storm behind the one being observed. Net control can help by relaying information to the storm spotter, keeping them updated on what else to expect. Net control should relay to the mobile storm spotters what is heading their way, any associated storm reports, changes in storm speed or direction, expected time of arrival, and perhaps even a recommended safe location to move to. Net control can do a lot to help keep spotters aware and safe during severe weather.

Spotting in Pairs

At no time should a storm spotter operate alone while mobile. It is best to spot in pairs with one focused on driving and the other on spotting. It is dangerous enough to drive near severe weather, even more so if distracted by trying to storm spot at the same time. While out spotting the driver not only has to deal with adverse road conditions but also be alert to others on the road, following traffic laws, and keeping aware of escape routes. The driver's only task should be driving; doing this will allow the spotter to focus attention on observing weather conditions. The spotter should also be the one who takes care of any equipment in the car such as radios, cameras, cell phones, and other gear.

Every effort should be made to minimize distractions to the driver. Both, or all in the vehicle, need to keep alert to potential hazards. Safety is everyone's responsibility.

Driving and Vehicles

The driver assumes a great deal of responsibility when out spotting. There are hazards on the road during severe weather that the driver must stay alert for. The driver must keep alert for emergency vehicles that will be on the road. Severe weather often calls for a greater response from public safety officials. They may be responding to a fire caused by lightning, power lines down across a road way, a

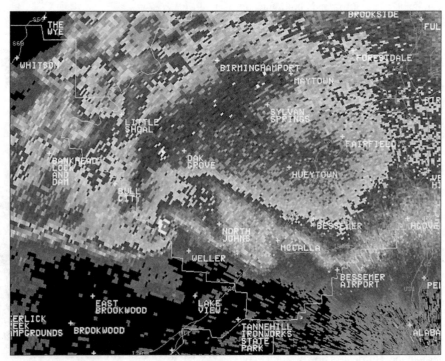

Tuscaloosa, Alabama April 2011. This is an image of the 0.5 degree reflectivity data from KBMX at 538 pm of the circulation as it crossed over the Tuscaloosa-Jefferson County line. [NOAA's National Weather Service, www.weather.gov/bmx/event_04272011tuscbirm]

2.2   Chapter 2

Southwest Winnebago County, Illinois, June 22, 2013, [Image ID: wea04582, NOAA's National Weather Service (NWS) Collection. Janice Thompson, photo]

submerged vehicle, or a traffic accident. Public safety responders must respond in a timely manner. When visibility is limited, it may be harder to hear an oncoming siren due to rain, wind, or noise from inside the vehicle — the driver may not be aware of an emergency vehicle as they would under normal conditions.

In addition to emergency vehicles, the driver must be aware of other traffic on the road — including pedestrians. Remember that visibility may be limited and this can affect reaction time.

Severe weather can also create road hazards that may not normally be encountered. Flood waters can wash away roads and bridges; ice and snow can cause roadways to be slick and hazardous; and strong winds can blow a vehicle off the road. It is also advisable to avoid gravel and dirt roads if there has been intense rain or flooding. These roads may become soft, causing vehicles to get stuck. It is better to observe from paved roads and better still from four lane intersections. These will provide the greatest number of possible escape routes on good roads.

Along with being aware of what is on the road, the driver must remember to observe all the rules of the road. It is too easy to get in a hurry when storm spotting. You should drive at speeds appropriate for road conditions and never in excess of the posted speed limit. And everyone in the vehicle must wear their seat belts.

The driver should also never attempt to drive through the core of the storm. Storm chasers call this punching the core. By driving through the core of the storm you run the risk of encountering deadly hail, intense rain, and if there is a tornado present you may not know it until it's too late.

If you do have to stop the vehicle there are also certain safety precautions that should be taken. First is to make absolutely certain that you are out of the travel portion of the roadway. While the shoulder of the road is usually a place where a vehicle can pull off, it can be a dangerous spot if visibility is limited. Try to find someplace such as a parking lot or rest stop area where you will be out of the way of traffic. Also avoid trees, power lines, and large signs. Strong winds can knock these down and pose a hazard to anyone nearby. Do not take cover under a highway overpass. Strong winds or tornadoes can send debris under the overpass and can result in injury or death. If conditions get too dangerous, get inside a sturdy structure.

As mentioned before, spotting at night is far more risky than spotting during daylight hours. Visibility is of course lower due to low light. This may make it difficult to see what is coming your way. Lightning may only give you momentary glances of the storm, but at the same time can also cause brief flash blindness. Not being able to see the storm will also make it more difficult to maneuver around the storm.

In addition to driving carefully, avoiding road hazards and conditions, and locating a safe spot and escape route, you must also make sure your vehicle is maintained properly. Not taking simple maintenance steps may pose safety hazards when out spotting. One of the most important things is making sure you can see what's coming at you when it's raining. Make sure wiper blades are good and applying something like *Rain-X* can help keep the windshield clear. Your vehicle needs to be in good mechanical condition. Make sure all fluids are filled to the appropriate level, tires properly inflated, and that all the lights work. And you need to make sure you have enough fuel. As a general rule you shouldn't head out spotting with less than half a tank of gas. Keep in mind that you may not be taking a direct route to where you are going. And if roads become flooded or blocked there may be many detours getting back. And while out spotting keep the vehicle's engine running when stopped as a precaution.

There are also things that you can keep in your vehicle to help you stay safe while out storm spotting. It should go without saying that amateur radio operators will have communications equipment in the vehicle with them, but don't forget extra batteries for the handheld transceiver, the cell phone and car charger, a list of important phone numbers, and a repeater directory or database app such as RFinder.[2] Also be sure to keep a basic first aid kit in the vehicle as well as a fire extinguisher. Have a blanket available in the vehicle. In winter weather it can be used to keep warm if you experience car trouble and in the event of a hail storm it can be used to provide a little additional protection. Make certain that you have a flash light and batteries in the vehicle too. And don't forget to have NOAA Weather Radio with you.

When you are storm spotting mobile you have to exercise extreme caution and be prepared for all hazards that you may face. A vehicle provides less shelter than a structure but this can be compensated for, to a certain degree, by keeping distance between you and the storm, planning an escape route, spotting in pairs, and making sure you and the vehicle are prepared for spotting. For further reading on the subject of mobile safety during severe weather see *Storm Chasing with Safety, Courtesy, and Responsibility* by Charles Doswell III of the National Severe Storms Laboratory.[2] Although written by a storm chaser, it has many valuable lessons for storm spotters.

SPOTTING FROM A FIXED LOCATION

Some storm spotters will report from a fixed location such as a home, emergency operations center (EOC), fire station, or shelter. Regardless of where they are located there are also certain safety precautions they should take during severe weather. Fixed locations offer a greater degree of shelter than a vehicle. There is a trade-off though. When in a vehicle you can leave a specific area quickly if it suddenly becomes hazardous. When in a home or other fixed location you must keep aware of what is heading your way. If you have enough time you may be able to move to a safer location. But if the weather changes suddenly you may have to stay put and take cover. The fixed location storm spotter should have a plan for dealing with each potential weather hazard. There are several safety considerations the fixed location spotter must keep in mind.

The Location

The first thing the spotter must consider is the location itself. Before severe weather strikes the spotter should do an assessment of their location to determine weaknesses and strengths. At what point will wind damage the structure? Is it in a flood plain? What is the water table like? What damage is possible from falling trees or power lines? We assess our location based on the likely threats we will encounter. If I reside in the Midwest I will likely assess the impact of threats such as tornadoes, floods, and winter storms. The threats you will face depends upon where you are.

Along with the structural concerns amateur radio operator storm spotters must also consider antennas and towers. Whether we put up a tower and antennas or string antennas up from trees and other supports, we must keep in mind how severe weather will impact them. Strong winds can take down towers and antennas so you will need to make sure they don't risk hitting power lines or structures. Ice can accumulate and damage antennas and

Lightning over National Severe Storms Laboratory mobile mesonet. Enid, Oklahoma, May 15, 2009. [J. J. Gourley, NOAA/NSSL, photo].

The tornadic storm observed earlier was accompanied by large dangerous hail which smashed the windshield of this National Severe Storms Laboratory instrumented vehicle. Note leaves and mud on side of truck. LaGrange, Wyoming, June 5, 2009. [VORTEX II, photo]

The National Weather Service forecast center in Norton, Massachusetts. [National Weather Service Boston/Norton photo]

towers. Flooding can cause ground to break away. And a lightning strike can not only destroy towers and antennas but also strike the house.

Operating

Staying on the air during severe weather can be a tremendous challenge. For the storm spotter, staying on the air, online, or on the phone to submit storm reports is essential. But staying operational during severe weather also requires awareness of certain safety issues. The greatest safety issue we face operating during severe weather is lightning. Lightning gives no warning that it is going to strike a specific location. But there are precautions that can be taken to minimize the damage caused by a lightning strike. Refer to Appendix 8 for a guide on lightning protection for the amateur radio station. Here are a few safety tips to follow when lightning is present while active as a storm spotter from home.

• Stay off of corded telephones. Lightning can travel along telephone lines. If you need to phone in a weather report use a cell phone.

• Use a handheld radio. Even with the best lightning protection it is best not to tempt fate. Avoid using a radio that is connected to an outside antenna.

• Avoid items that can conduct electricity. Lightning can enter the house from many sources; wires, pipes, even the ground. It does not take a direct hit for lightning to get into the home.

Being inside a house is one of the best steps you can take for lightning protection. It does not mean you are in a lightning-proof place, though. Read and follow the lightning safety guidelines available from the NWS.

Operating From an EOC or Other Facility

Often during SKYWARN activation, and ARES activation for that matter, net control will be located at an emergency operations center (EOC). There are some distinct advantages to this. It allows net control to stay in touch with critical players in emergency response, such as emergency management, public safety, public works, and government officials. This close contact allows important weather information to be disseminated quickly to key individuals. Operating from an EOC also allows a dedicated station to be set up for emergency operations. Net control also may operate from fire stations, hospitals, shelters, or other similar locations. The general idea for all of these is to keep net control close to important decision makers and make it possible to relay important information quickly.

For our purposes we will assume net control is operating from an established EOC. The safety guidelines for our EOC net control will be similar to those that net control stations at other locations should follow.

Operating net control at an EOC is a lot like operating from a public safety dispatch center. It is a dedicated station. It operates during emergencies and crises. It is manned by an operator. And it has to stay on the air. To do this safely the net control operator is going to want to find out a few things. First, is that station grounded? Is there lightning protection? What kind of power backup is there? Generator? Batteries? What station equipment is connected to the backup power system? Is there an evacuation plan and is it posted? Is there a first aid kit handy? Is there backup amateur radio equipment? The net control operator must know the answers to these questions.

Just because we're located at a station in a secure facility and designed to stay on the air doesn't mean we are immune

The impact of flooding should not be underestimated. Flood waters at the home of George, KK4F. [George Mudd, KK4F, photo]

from the effects of severe weather. Lightning can still strike and cause injury or death and damage equipment. The EOC can be in the path of the tornado. In 2003 an EOC in central Indiana was inundated with flood water because it was located in a basement. Amateurs who volunteer as net control in EOCs and similar facilities should be familiar with potential hazards. Talk with the EOC manager or other official in charge about any risks the center faces. Talk with other amateurs that have operated as net control from that EOC and learn about their experiences. Just because it is an emergency operations center does not mean it is secure. Many EOCs have been placed in less-than-ideal spots as ways for jurisdictions to save money or as the result of poor planning.

NWS WEATHER SAFETY INFORMATION

For decades the NWS has produced information in the form of pamphlets and flyers, videos, and websites. In its actual severe weather monitoring, NWS produces warning and advisory information. Safety information is also included on NOAA Weather Radio broadcasts. Weather safety information is one of the most important NWS products.

Currently the NWS produces safety information on lightning, hurricanes, thunderstorms, floods, tornadoes, heat, and winter weather, just to name a few. Every SKYWARN storm spotter should be familiar with severe weather safety. We cannot get into the frame of thought that we know enough just because we took the SKYWARN training. Read and heed the safety guidelines from the NWS (see the sidebar, "NWS Weather Safety Information"). No storm spotter should *ever* compromise safety to turn in a weather report.

NWS Weather Safety Information

The following weather safety information has been provided by the NWS for this publication and it is used with permission of the National Weather Service

Tornado Safety
- Be on the lookout for other tornadoes that could form in the vicinity of the tornado you are watching.
- Never try to outrun a tornado in an urban or congested area, but immediately get into a sturdy structure after parking your car out of the traffic flow.
- Do not take shelter under bridges or overpasses. These structures do not offer ample protection and could increase the chance of injury or death.
- If you are caught outdoors, seek shelter in a basement, shelter, or sturdy building. If you cannot quickly walk to a shelter, immediately get into a vehicle, buckle your seat belt and try to drive to the closest sturdy shelter. If flying debris occurs while you are driving, pull over and park. Now you have the following options as a last resort:
  – Stay in the car with the seat belt on. Put your head down below the windows, covering with your hands and a blanket if possible.
  – If you can safely get noticeably lower than the level of the roadway, exit your car, and lie in that area, covering your head with your hands.
  – Your choice should be driven by your specific circumstances.
- Flying and falling debris is the biggest hazard in a tornado. To be safe, you should get in, get down, and cover up. Underground or in a safe room is best. If no underground shelter is available, get to the center of a sturdy building on the lowest level. Put as many walls between you and the tornado as possible. Stay away from windows and doors. Cover up to help minimize being injured by flying or falling debris.

Flash Flood Safety
"Turn Around, Don't Drown!"
- Do not attempt to drive or walk across a flooded road or low water crossing. You cannot be sure about the depth of the water or the condition of the roadway — it might be washed out.
- Two feet of moving water will carry away most vehicles.
- Six inches of fast-moving water can knock you off your feet.
- If your vehicle is suddenly caught in rising water, leave it immediately and get to higher ground.
- Be especially careful at night when flash floods are harder to recognize.

Lightning Safety
"When Thunder Roars, Go Indoors!"
- Remain in a hard-topped vehicle or an indoor location for at least 30 minutes after you hear the last thunderclap. If you use radio equipment, avoid contact with it or other metal inside your vehicle to minimize the impacts should lightning strike.
- If you are out on the water and skies are threatening, get back to land and find a fully enclosed building or hard-topped vehicle. Boats with cabins offer a safer, but not perfect, environment. Safety is increased further if the boat has a properly installed lightning protection system. If you are inside the cabin, stay away from metal and all electrical components.
- When reporting to the NWS, use a cordless or cell phone if available.
- Lightning victims do not carry an electrical charge, are safe to touch, and need urgent medical attention. If a person has stopped breathing, call 9-1-1 and begin CPR if the victim is not breathing.

Downburst Wind Safety
- Keep a firm grip on your vehicle's steering wheel to maintain control. Downbursts can occur suddenly with an abrupt change in wind speed and direction.
- If you can do so safely, point your vehicle into the wind to minimize the risk of the vehicle being blown over.
- Be prepared for sudden reductions of visibility due to blowing dust or heavy rain associated with downbursts.
- Spotters observing from a substantial building should move away from windows as the downburst approaches.

Hail Safety
- Substantial structures, such as a garage, offer the best protection from hail.
- Spotters in vehicles should avoid those parts of the storm where large hail is occurring.
- Hard-top vehicles offer good protection from hail up to about golf ball size. Larger hail stones will damage windshields.

SPOTTER SAFETY OFFICER

Training and severe weather awareness can do a lot to help keep us safe during SKYWARN activation. However, when we're out storm spotting or operating at the net control station site we can easily get so caught up in the many tasks that have to be done that we may overlook safety issues. We can borrow an idea from industry and public safety — the safety officer, or for our purposes the Spotter Safety Officer (SSO).

The SSO is appointed by either the local SKYWARN coordinator or net control station (NCS). Their duty is to keep constant tabs on storm spotter positions relevant to the storm. If conditions dictate, the SSO will advise the NCS to have spotters move to a new location. To do this successfully the SSO must carefully monitor the storm location using radar and information from the NWS and local news media, know where spotters are located, pay careful attention to spotter reports of hazardous weather in the area, and plot this data on a map. Because safety is at stake the SSO must be knowledgeable about severe weather, storm structure, and storm characteristics. The individual selected for SSO must be someone that spotters and net control trust to look after their safety. It is not a position to be taken lightly.

The success of the SSO is dependent on the storm spotters. Spotters must go to their assigned locations and only move at the direction of net control, when safety dictates, or when activation is ended.

A final duty of the SSO is to advise the net control station or local SKYWARN coordinator on safety concerns. If an individual storm spotter is acting in an unsafe manner the SSO should advise the NCS or the SKYWARN coordinator, who can then take action.

In the same way amateur radio polices itself, so too should those involved in SKYWARN activities. There is no room for unsafe behavior.

REFERENCES

[1]M. Mitchell, "Indiana Tornado," *9-1-1 Magazine*, May 2006.

[2]The *ARRL Repeater Directory* book is published annually. RFinder is available as an iOS or Android app with an annual subscription fee. Information on both products is available from **www.arrl.org/shop**.

[2]Charles A. Doswell III, "Storm Chasing with Safety, Courtesy and Responsibility," **www.flame.org/~cdoswell/chasesums/Chase_safety.html**

Equipment and Resources

When the local SKYWARN group is activated during severe weather, many individuals come together for a common purpose — to observe and report severe weather conditions to the National Weather Service and other served agencies. To do this job, though, requires the right tools. Whether you are a home-based spotter, a net control station, or a mobile storm spotter, having the right equipment can make you more effective.

In this chapter we will look at the wide range of gear a storm spotter may use. We will look at mobile and home station radios, antennas, GPS, lights, "go-kits," weather equipment, and other gear that can be useful. We will go beyond the radio to see what other items may be helpful when storm spotting. We will also look at apps and online resources for the storm spotter and see how digital modes can be useful.

So let's get started with the most important items: radios and other communication devices.

This SKYWARN net control station is located at a volunteer fire department. Left to righ,: Caleb McCormick monitoring the NWS website; Larry Brown, K5LMB, handling check-ins on the SKYWARN 2 meter net. [Mike Corey, KI1U, photo]

RADIOS AND ANTENNAS

For the amateur who is storm spotting, the radio is probably the most essential piece of equipment. Whether he or she is acting as net control, at home, or in the car, the radio is the fail-safe link to call in reports of severe weather.

There are many factors that need to be considered when deciding which radio will work best. Let's look at the considerations for home-based spotters, net control stations, and mobile spotters.

Which radio to use of course depends on what role you are taking as a spotter and what frequencies and modes you will need to use. Most local SKYWARN nets are conducted on local VHF or UHF repeaters. A mobile VHF/UHF radio is a good choice whether you are a mobile spotter, net control, or a home-based spotter. Current mobile radios come in a wide range of options: single band, dual band with single or dual receive, radios that

This is KL7FWX, the SKYWARN amateur radio station at NWS Fairbanks, Alaska. [National Weather Service photo]

Equipment and Resources 3.1

provide digital voice modes such as DMR, D-STAR or C4FM (Yaesu System Fusion), and radios that combine HF/VHF/UHF in one package. Most operate on 13.8 V dc, so they are easy to connect to a power supply or battery.

Handheld radios, while convenient, are not always the best for storm spotting. They are dependent on battery life and have a lower operating power. The stock flexible antenna is less efficient than a full-size antenna, and the audio output of the speaker may not be high enough to hear over ambient noise. Handhelds can be useful to the spotter though. Having one in the car can give the mobile spotter an extra receiver that can be tuned to the local NOAA weather channel, used as a backup radio, or used to monitor local public safety channels (if legal to do so in your area). Net control operators may use a handheld to monitor traffic if they have to step away from the radio. A handheld radio is also useful when lightning is present because it allows the operator to switch to a radio not connected to an outside antenna, thereby reducing risk of injury.

There are times that an HF radio will be needed for storm spotting activities. Local nets are used to relay information to National Weather Service offices. This can be done by telephone, internet, or radio. The local NWS office may have an amateur radio station set up to run an HF SKYWARN net to collect reports from local nets in their area. If you are a net control station, relay station, or a home-based station, you may need HF capability to send reports to the NWS office. In some cases, mobile storm spotters may need to utilize HF to relay reports to net control. Additionally, HF can be used to link multiple NWS offices. This is particularly useful during hurricane operations.

Mobile setup for 2-meter voice and APRS. [Mike Corey, KI1U, photo]

The antennas we use for storm spotting are not too different from those normally used in amateur radio. There are certain considerations that must be made by the storm spotter when selecting antennas, though.

For the mobile storm spotter, the two biggest factors will be frequency coverage and gain. Most mobile spotters will likely use VHF/UHF repeaters to radio in weather reports and they will do so from their local area. Most mobile VHF/UHF antenna setups will be able to do this without a problem. But if your local repeaters coverage is limited, you may want to look at a higher gain antenna.

For effective home stations that may be serving as relays or net control, there are two key factors for antennas: survivability and being heard. We can learn a lot about antenna survivability from the contest community and DXpeditions. These communities know the importance of staying on the air. A close study of station building techniques of contest stations will serve the home-based spotter well. DXpeditions often set up in remote, far-off locations. Antennas have to be lightweight, resilient, and effective to get a signal out. Gain is another critical factor on antenna choice, especially if you are running low power. HF directional antennas are a tremendous help for regional communications, but they may be susceptible to damage from winds and ice. It is a good practice to keep simple wire HF antennas ready to deploy in case of emergency.

NOAA WEATHER RADIO

The roots of NOAA's Weather Radio service (NWR, www.weather.gov/nwr/) go back to 1906, when the US Weather Bureau experimented with radiotelegraphy to speed notice of weather conditions. From the 1950s onward it was designed to broadcast weather information to the general public, aviation, and marine communities. It included forecasts, daily observations, watches, and warnings. In the 1990s the mission of NWR was expanded to include broadcasts about all hazards. Today NWR has more than 1000 transmitters, covering all 50 states, adjacent coastal waters, Puerto Rico, the US Virgin Islands, and the US Pacific Territories. In conjunction with federal, state, and local emergency managers and other public officials, NWR also broadcasts warning and post-event information for all types of hazards, including natural (such as earthquakes or avalanches), environmental (such as chemical releases or oil spills), and public safety (such as AMBER alerts or 911 telephone outages). If you tune to the local NWR frequency, though, most days you will get a steady stream of weather-related information.

Today we have a wide variety of sources to receive weather information, so why is NWR still needed? A January 1975 White House policy statement that remains in effect designated NOAA Weather Radio as the sole government-operated radio system to provide direct warnings into private homes for both natural disasters and nuclear attack.[1] NWR also remains the primary method for activating our nation's Emergency Alert System (EAS) in which emergency messages from the government are relayed to the American public through a variety of radio and television broadcast means.

Even though watches and warnings can be received from the NWS through the

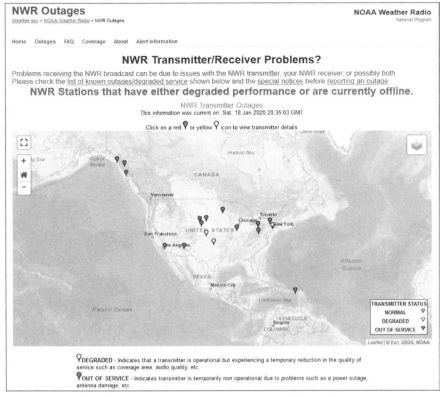

NOAA Weather Radio publishes an online Outage Map and other information about stations that are not operating with normal capabilities at www.weather.gov/nwr/outages.

NOAA Weather Radio on Broadcastify. [Screen capture courtesy Broadcastify]

internet and smartphone apps, text messaging services, and email, NWR still provides a valuable service. Twenty-four hours a day, seven days a week, you can receive constant, reliable weather information. And when a watch or warning is issued NWR has about a two minute lead time over other media. This kind of heads-up can be critical when severe weather threatens. Similar to a smoke detector, an NWR broadcast can wake you in the middle of the night to alert you to a dangerous situation. In addition, NWR coverage is almost 100% throughout the United States and its territories.

Reception of NWR does require a special receiver. There are commercially available NWR receivers that can tune to any one of NWR's seven frequencies (see **Table 3.1**). Many of these have battery backup and Specific Area Message Encoding (SAME). SAME is a digital tone sent over normal audio using AFSK and carries an alert message for a specific area.

SAME areas are generally organized by county, parish, city, or maritime area. So if you are setting up a NWR receiver and you live in Oxford, Mississippi, you will set the frequency to 162.55 MHz and program the SAME code 028071 (Lafayette County). To find your local NWR frequency and SAME code you can go to the NWR website.

For the best performing NWR receivers, NWS suggests that you look at devices that have been certified to Public Alert standards. These radios meet certain technical standards and come with many features such as SAME, battery backup, both audio and visual alarms, selective programming for the types of hazards you want to be warned about, and the ability to activate external alarm devices for people with disabilities.

Reception of NWR is not limited to

Table 3.1 Frequencies Used by NWR

162.400 MHz
162.425 MHz
162.450 MHz
162.475 MHz
162.500 MHz
162.525 MHz
162.550 MHz

Breakdown of a SAME Message

1) **Preamble and Header Code.** The preamble and header code are transmitted three times with a one second pause (±5%) between each coded message burst prior to the broadcast of the actual voice message. Then the preamble and EOM code are transmitted three times with a one second pause (± 5%) between each EOM burst.

2) **Warning Alarm Tone/Attention Signal.** The Warning Alarm Tone (WAT), if transmitted, is sent within one to three seconds following the third header code burst. The frequency of the WAT is 1050 Hz (± 0.3%) for 8 to 10 seconds at no less than 80% modulation (± 4.0 kHz deviation minimum, ± 5.0 kHz deviation maximum).

3) **Voice Message.** If transmitted, the actual voiced message begins within three to five seconds following the last NWR SAME code burst or WAT, whichever is last. The voice audio ranges between 20% modulation (± 1 kHz deviation) and 90% modulation (± 4.5 kHz deviation) with occasional lulls near zero and peaks as high as, but not exceeding, 100% modulation (± 5.0 kHz deviation). Total length of the voice message should not exceed two minutes.

4) **Preamble.** Repeat previous preamble.

5) **End Of Message (EOM).** EOM is identified by the use of "NNNN."

For more information, see **www.weather.gov/nwr/nwrsame**.

weather radios, however. Most VHF/UHF transceivers are capable of receiving NWR. Several have NWR as a one-touch feature with all NWR frequencies stored in a separate dedicated memory bank. While these radios may not be able to be programmed with the SAME codes, many are capable of picking up the 1050 Hz tone that is generated when there is an emergency alert. The only difference is that this is not an area-specific tone. Additionally you can access some NWR stations on apps such as Broadcastify.

NWR is a valuable tool for the storm spotter. The net control station should be monitoring NWR for up-to-date weather conditions, watches, and warnings. Home based spotters and mobile spotters should also be able to monitor NWR. A handheld VHF/UHF radio tuned to NWR is a handy tool to have when mobile spotting since the flexible antenna will be able to pick up the nearest NWR transmitter in most areas. It is also important to be aware of issues with your local NWR repeater site. Information on outages and degraded service can be found on the NWR website. Users can also report outages through the website.

DIGITAL MODES FOR THE STORM SPOTTER

APRS

Developed by Bob Bruninga, WB4APR, APRS — the Automatic Packet Reporting System — is "a two-way tactical real-time digital communications system between all assets in a network sharing information about everything going on in the local area."[2] It provides amateurs with the ability to share a wide range of data not only at a local level, but by use of internet gateways at the global level as well. APRS has several important features for storm spotting. Of primary interest to the storm spotter are position information, weather data, and messaging.

Probably the most widely known feature of APRS is the ability to transmit position reports and see these reports on a map. This feature is valuable for storm spotters, net control stations, NWS offices, and emergency management during severe weather. By logging into one of the many online sites that provide APRS information (such as **aprs.fi**), you can quickly see who is currently on, their location, and type of station. Net control, emergency management, and NWS can see if there are mobile spotters out and, if need be, reposition them or determine quickly if they are in harm's way of approaching severe weather.

Weather data can also be distributed via the APRS network. Using a program such as *UI-View*, weather data from local amateur radio APRS stations and CWOP stations (Citizen Weather Observer Program, discussed later in this chapter) can be seen, along with NWS watches and warnings. Mobile stations can see data being transmitted from weather stations and receive information about weather watches, warnings, and advisories.

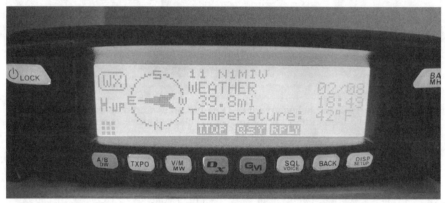
Weather information received via APRS. [Mike Corey, KI1U, photo]

APRS also allows users to transmit short text messages to others. They can be sent from one APRS user to another, an APRS user to an email address, or even to a cell phone as an SMS text message. The Amateur Radio Universal Text Messaging/Contact Initiative is an effort underway to greatly expand the ability to send messages to users over a wide range of systems. While text messaging via APRS has been around for many years, there are still a lot of other systems that can be integrated into a single messaging system. This is a key feature of APRS that spotters should be utilizing and helping to advance and develop.

While APRS offers much more than position, weather, and messaging, these are critical components for storm spotters. If you have not tried APRS yet give it a try! For more information on APRS and its capabilities refer to the *Amateur Radio Mobile Handbook* and *Radios to Go!*.[3, 4]

EchoLink

EchoLink has been a valuable tool to the SKYWARN community. We won't go over all the details on what EchoLink is here, but instead will focus on how it can be utilized by amateur radio storm spotters. If you are not familiar with EchoLink or other VoIP communications for amateur radio, refer to the EchoLink website or *VoIP: Internet Linking for Radio Amateurs* by Jonathan Taylor, K1RFD.[5, 6]

EchoLink and other VoIP modes such as IRLP (Internet Radio Linking Project) allow us to combine RF communications and the communications capability of the internet to form a more robust network. By utilizing EchoLink, a SKYWARN net control station can keep in touch with all local SKYWARN nets in a particular county warning area. Areas that do not have repeater coverage can still stay in touch by using EchoLink in an internet-only mode connecting directly with the SKYWARN net control station at the

NWS office. Utilizing EchoLink in this way can help fill in any communications gaps and generally make severe weather communications more effective.

Let's take a look at how EchoLink is used in a real situation. At the Tampa, Florida, NWS office amateur radio is a critical part of severe weather operations. SKYWARN nets, activated during severe weather outbreaks, can be conducted at the local level using local repeaters or at a regional level using the NI4CE repeater system which has an EchoLink node. During regional weather events such as hurricanes, the EchoLink node can help provide additional communications capabilities by filling in repeater coverage gaps. As long as stations can access the internet, they can still communicate with net control. It also provides a way for others outside the range of local repeaters to access net control. For example, let's say the WX4NHC at the National Hurricane Center needs to get in touch with SKYWARN net control at Tampa WFO (Weather Forecast Office). Utilizing the EchoLink connection they can do so with only an internet connection even though the net control station in Tampa may have no internet connection and is operating through the repeater.

An added benefit to having EchoLink is during the post severe weather phase. EchoLink can be used as a means to conduct a regional net. Stations checking in can share reports of events that occurred in their area, discuss problems encountered and possible solutions, and make reports to district and state emergency coordinators. In other words, it can be a conference call tool, post-event.

A great example of this is the NWS Ruskin SKYWARN Practice and Outreach Net. This net meets each Tuesday evening at 9 PM local time on the NI4CE repeater. The net control station operates from the NWS office. The purpose of the net is "to provide limited SKYWARN training, make announcements of interest to SKYWARN personnel, provide local weather information (especially if severe weather is expected in the next few days), and to provide net participants the opportunity to interact with the staff of their local NWS weather forecast office." By combining the repeater system and the EchoLink node a much wider audience can be reached.

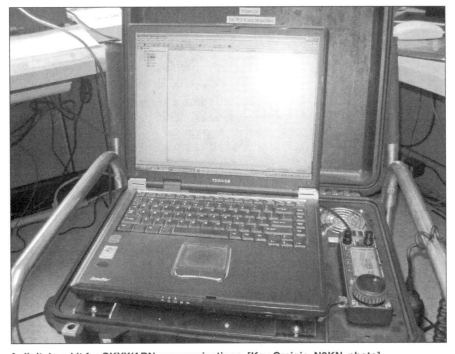

A digital go-kit for SKYWARN communications. [Kay Craigie, N3KN, photo]

EchoLink and IRLP During Severe Weather

There are a number of ways EchoLink and the Internet Radio Linking Project (IRLP) can be utilized during severe weather and significant winter weather situations. Let's look at some examples of this activity via the operations at WX1BOX, the amateur radio station at the National Weather Service in Norton, Massachusetts.

Similar to how the VoIP Hurricane Net utilizes a network that allows for an EchoLink conference and an Internet Radio Linking Project or IRLP reflector channel to be connected, the New England reflector system is utilized by WX1BOX during severe weather events. It combines EchoLink conference *NEW-ENG3* node: 9123 with IRLP 9123. This network is utilized to gather reports from around New England. At times, the National Weather Service in Gray, Maine, WX1GYX, is also on the network and has gathered reports from their coverage area on the network. This network allows for repeaters both on EchoLink or IRLP, other IRLP and EchoLink links that are on simplex and EchoLink PC and phone users connect to the network providing a regional weather net with this capability. The regional weather network concept can be utilized for any weather hazard from severe thunderstorms and tornadoes to flooding to significant winter weather and high wind storm events.

EchoLink and IRLP can also be utilized to reach local nets that don't typically connect up to regional nets and may be out of RF range for a NWS Forecast Office. For example, to reach the Hartford-Tolland County SKYWARN Net in the WX1BOX NWS Boston/Norton coverage area, WX1BOX can either connect in via EchoLink on a pc or EchoLink phone app or via IRLP node. Operators at WX1BOX have used all of those means to connect to this SKYWARN Net to collect damage and meteorological data when this part of the NWS Boston/Norton coverage is affected by severe weather. If internet fails, a liaison from this net will go to another area repeater via RF, utilize VHF simplex or HF to reach WX1BOX to allow for the reports to continue to flow in this situation.

VHF Digital Voice

Over the last couple of decades, amateurs have developed or made use of new communication modes such as APRS, FT8/FT4, PSK31, Winlink, and a variety of digital voice modes. Some of these modes can be useful tools for the amateur radio operator who is a storm spotter. We can put them to use to make our ability to communicate during severe weather more robust. Digital voice modes on VHF/UHF include D-STAR, DMR, C4FM, and P25. Typically these modes utilize internet linking to create an incredibly large network. Storm spotting doesn't always need to tie up large networks, but some modes can be customized to create large area or regional networks that allow multiple SKYWARN nets to communicate with one another and the NWS forecast office.

There are a few things we must keep in mind before pursuing digital voice options

D-STAR and Severe Weather

By John Davis, WB4QDX, Assistant Section Emergency Coordinator, Georgia

Please briefly describe what *D-STAR* is.
D-STAR, or Digital Smart Technology for Amateur Radio, is an open protocol created by the Japan Amateur Radio League for transmission of digital voice and data. The D-STAR protocol allows digital voice and simultaneous low-speed data on the 2 meter, 70 centimeter, and 23 centimeter bands. In addition, it defines a high-speed data standard (128 kbps) in the 23 centimeter (1.2 GHz) band.

How many D-STAR repeaters are there? How can I find out if there is one near me?
There are thousands of registered D-STAR users around the world and hundreds of registered D-STAR systems. Each D-STAR system may consist of up to four separate repeaters operated by one repeater controller, so the number is well over 1000 individual repeaters operating in the three amateur bands (2 m, 70 cm and 23 cm). Two websites are useful for locating D-STAR repeaters near you. They are www.dstarusers.org and www.dstarinfo.com. The www.dstaruser.org website has a "Last Heard" page that lists users who have accessed D-STAR systems in the last 24 hours.

What equipment do I need to get started on D-STAR?
Currently, Icom and Kenwood sell D-STAR compatible radios. In addition, the DV Dongle is available to connect to your PC's USB port and uses the sound card to access D-STAR repeaters around the world similar to EchoLink.

Are SKYWARN or any other severe weather nets run on D-STAR?
There are several nets now operating on D-STAR. One specifically for severe weather is the Southeastern D-STAR Weather Net. It meets each Sunday night at 9 PM ET as a training and practice net and will operate during times of severe weather in the Southeastern US, an area subject to tropical weather, hurricanes, tornadoes and severe thunderstorms. Using D-STAR reflectors, which are like conference bridges, multiple repeaters and DV Dongle users can connect together for a net covering a wide area with consistent voice quality across the net and without the beeps and varying levels of traditional linked systems. An advantage of D-STAR for severe weather reporting is that repeaters and areas can be linked together quickly in whatever configuration is needed.

Are there any differences in how a net is run on a D-STAR repeater over a conventional repeater?
There doesn't have to be, but D-STAR does bring some unique capabilities that are advantageous for nets. Since the user's call sign is sent with each transmission, each user is identified without giving the call sign verbally. The Southeastern D-STAR Weather Net uses this capability for quick check-ins using the "Quick Key" format. When the Net Control Station calls for check-ins, users can just key their radio for one second and their call sign is visible to the net control. This allows a list of users to check-in in a shorter period of time. It can also be used for breaking stations by just keying between stations. The other stations see the new call sign displayed on their radio and can recognize the breaking station.

Also, the simultaneous low-speed data capabilities of D-STAR allow "packet speed" data to be sent with voice transmissions. It also allows for separate data to be sent using the same repeater. Comparable FM systems may require a separate radio and TNC from the voice radio for SKYWARN. First, do we need that level of complex network capability, or is a conventional repeater appropriate? Second, by switching over to digital voice, does it leave out some of our SKYWARN spotters who do not have digital voice radios? Third, digital voice typically has less audio fidelity than FM, so we must assess if critical messages can be heard and understood.

One of these modes, D-STAR, has been used for severe weather and hurricane response in the southeast United States for several years. In the accompanying sidebar John Davis, WB4QDX, explains how D-STAR works with severe weather and the Southeast D-STAR Weather Net.

In recent years DMR has risen in popularity around the country. Use of DMR for SKYWARN is emerging, but still not as widespread as analog FM. NWS WFO Gray, Maine has been making use of DMR as a means to add coverage for areas not in the footprint of an analog FM repeater. The availability of DMR radios and the support of the New England Digital Emergency Communications Network (NEDECN) have added capacity to the amateur radio coverage of their CWA. DMR is now the second most popular means for turning in SKYWARN reports, and NEDCN has even added a SKYWARN talk group.

along with separate repeaters or digipeaters to handle voice and data.

Are any of the National Weather Service offices on D-STAR?
I don't have information on NWS offices around the country, but our plan for Georgia includes providing one in the NWS office in Peachtree City and other offices serving Georgia. Peachtree City has a well equipped and active amateur radio station right with the meteorologists and providing statewide capabilities for receiving incoming reports will be valuable. Also, with the D-STAR network already in place in neighboring Alabama and Florida, reports of approaching severe weather across state lines will be possible.

Please describe the severe weather net procedures on the D-STAR system.
The net is still very new and procedures are being developed as we go. The primary purpose of the net is to connect affected areas with resources outside of the area. D-STAR's versatile linking through reflectors makes connecting repeaters to form ad hoc networks a quick and easy process. D-STAR doesn't replace other amateur radio resources, but is another tool in the bag bringing new capabilities to support emergency events.

Since D-STAR is a network, is it more suited for larger events such as hurricanes instead of smaller isolated events such as tornadoes?
Actually, the voice and data capabilities of D-STAR make it equally suited for events over large areas and within a smaller, defined area such as tornado or other local event. In its simplest form D-STAR is a local repeater similar to a traditional analog repeater, but with the clarity of digital voice. D-STAR's low and high speed data capabilities over a single, or Gateway equipped, repeater can provide digital messaging to supplement voice communications or internet connectivity within an area where traditional infrastructure may be overloaded or unavailable. During a Gwinnett County, (Georgia) SET, D-STAR was used to relay photos from field locations directly to the EOC via the internet.

Can D-STAR support tracking and messaging capabilities like APRS?
Yes. All D-STAR radios are capable of sending GPS location information or messaging. The position information is transmitted in a format called DPRS. D-STAR repeaters gate the information to the internet. Unlike FM systems requiring a radio for voice and a separate radio and TNC for data or APRS, D-STAR transmits simultaneous voice and data in a single radio. With free messaging applications such as D-RATS, advanced messaging, and mapping data transmission capabilities are possible over D-STAR.

What advantages are there in using D-STAR over a conventional repeater for severe weather?
Both provide valuable services for severe weather reporting, but D-STAR's linking, data transmission and position reporting expands reporting capabilities. The clarity of digital transmissions can also improve accuracy of reporting.

What disadvantages are there?
Many amateurs still have analog radios, although D-STAR equipped amateurs and the availability of repeaters has grown at a very rapid pace. The cost of D-STAR radios is still higher than their analog counterparts.

NAVIGATION AIDS
GPS

Today we find GPS capability nearly everywhere, with standalone GPS units, smartphones, and even built into our transceivers and vehicles. Originally designed with the military in mind, it is now a part of everyday life. GPS can be a valuable tool for the storm spotter, and when combined with other technologies, even more so. Of course the main purpose of GPS is to answer the questions "Where am I?" and "How do I get there?" A GPS unit can provide accurate time, elevation, speed, heading, and tracking capability as well.

We have moved past the days when a standalone GPS unit was something of a novelty. GPS is everywhere and each of us has our favorites when it comes to GPS navigation. True navigation apps such as Google Maps and MapQuest will help us get to a specific location, while others such as Waze can give us lots of information about the roadways that get us there. Most of the time we can use our favorite navigation app and it will suit our needs. There are times, though, that we may have to look for other solutions.

While serving as part of the response to hurricane Maria in Puerto Rico, a team of Red Cross volunteers needed to navigate their way to the Guajataca Dam. The roads were difficult to navigate because a landslide washed portions away. Even local guides struggled to find alternative routes. All of the volunteers had cell phones capable of giving access to Google Maps, but even though the GPS was working fine, there was no data signal to download the maps from. One of the volunteers had thought ahead and had installed an app called *Offline Maps & Navigation*. This app allows the user to download maps of individual states, territories, and other countries and use them in tandem with the phone's GPS. The maps provide detail down to street level and allow for navigation to a destination.

Storm spotters should be prepared for what to do when the GPS is working, but the network isn't there. Apps such as this one are one solution. Google Maps also

Screenshot of downtown San Juan, PR from *Offline Maps & Navigation*. [Mike Corey, KI1U, photo]

offers an option to download maps ahead of time so you don't use too much data while enroute.

Maps

While GPS is a handy tool for the storm spotter, it is a good idea to have a bit of redundancy when it comes to navigation. In addition to the redundancy factor we should keep in mind that most navigation units in vehicles are set up for the driver to see and use. During severe weather the driver should be paying attention to driving and not be distracted. Navigation is best left to a second spotter in the passenger seat. Remember, safety always comes first. A good set of maps is useful to have, whether you are a mobile storm spotter, fixed, or work as a net control station. But what kind of maps should you have? Why not just keep a road atlas handy?

There are several types of maps that a storm spotter can use. First is a local street map. Although most mobile storm spotting is done in areas outside of urban congestion, it is still a good idea to keep one handy. Keep in mind storm spotters also report on flooding conditions and urban areas are more prone to flooding than non-urban areas. A local street map will allow you to mark locations that are flooded and perhaps even pencil in a flooded area. City and county street maps are also handy for identifying escape routes when spotting while mobile. Remember that storm spotters serve their local community, so having good local maps is important.

Local topographic maps are also useful. By using the contour lines and elevations, a spotter can identify a good spot to observe from and identify areas that may be particularly hazardous during severe weather such as low-lying areas prone to flooding.

A state highway map is handy to have in order to reference weather in relation to nearby towns, counties, highways, and other points. A road atlas can also be useful for referencing weather to geographic locations. Another handy set of maps to have is a topographic/GPS atlas designed for a state or portion of a state. DeLorme offers atlases for all 50 states that provide a wealth of information. Also handy are aviation maps. These are commercially available and provide aviation data on a topographic map. They cover sections of the United States each based on a major airport such as Memphis, St Louis, or Atlanta.

Something else to keep in mind is that maps don't have to be on paper. Net control stations and fixed storm spotters can utilize online mapping resources. Resources such as Google Maps and Google Earth should be in every SKYWARN net control operator's toolbox.

Many programs feature a drawing tool that allows the user to draw lines, points, labels, polygons, or other shapes on the map. Often these can be saved for future reference. A user can use this feature to plot tornado reports, hurricane tracks, mobile storm spotter positions, or hail reports on a map. Using the polygon feature allows plotting areas of flooding. And unlike a paper map it only takes a click or two of the mouse to erase and start again.

Another useful feature is GPS interface capability. This is especially useful for mobile spotters. It can be used to plot location and route, and eliminates time spent trying to figure out where you are on the map. Many mapping programs also feature topographic maps and can provide weather data, both useful features.

Probably the most important considerations, though, will be your operating platform and operating system. You will need to make sure that the software you choose can operate from your device, whether it is a tablet, smartphone, laptop or desktop. Also check and make sure it is compatible with your operating system. There is software available for just about every modern platform and system.

A particularly useful piece of software for the storm spotter that deserves special mention is Google Earth.[7] The software is capable of running on multiple platforms and systems, including Windows, macOS, Linux, smartphones and tablets to name a few. The basic software is free and there are other versions with enhanced features available for a subscription fee. Since its inception Google Earth has become one of the most widely used mapping programs.

Google Earth has several built-in features that are of use to storm spotters, and a wide range of plug-ins that can be added. All plug-ins for Google Earth are in KML format. This format, which is based on XML, is the file that contains geographic data for use in Google Earth. Plug-ins in KML format may be written by individuals to add custom data to Google Earth. Many of the KML files are available in a gallery on the Google Earth website, while others are available from other sites for free or for a charge.

A great feature of Google Earth is that it can be integrated with other programs relatively easily. See Appendix 6 for a guide on how to integrate Google Earth, APRS, and NWS data.

CELL PHONES AND INSTANT MESSAGES

Today the go-to device for communications is the smartphone. Over the past several decades, cellular technology has advanced beyond the original concept of a mobile telephone. Today cellular phones are capable of voice, data, text, app support and imaging, both still and digital. Many of these features can be valuable to the storm spotter. Let's look at several of these features that the storm spotter may want to consider for his or her tool box.

Most smartphones can now receive Wireless Emergency Alerts (WEAs), made available through FEMA's Integrated Public Alert and Warning System (IPAWS) infrastructure.[8,9,10] WEAs can be sent by state and local public safety officials, the National Weather Service, the National Center for Missing and Exploited Children, and the President of the United States. WEAs look like text messages, but are designed to get your attention and alert you with a unique sound and vibration, both repeated twice. WEAs are no more than 90 characters, and will include the type and time of the alert, any action you should take, as well as the agency issuing the alert. WEAs are not affected by network congestion and will not disrupt texts, calls, or data sessions that are in progress. Mobile users are not charged for receiving WEAs and there is no need to subscribe. To ensure your device is WEA-capable, check with your service provider.

Text messaging is a great way to exchange short messages between individuals and groups. Many schools and businesses use instant messaging to send alerts to students and employees. This feature can also be useful for severe weather preparedness. Notice of severe weather watches and warnings can be sent to a large number of recipients in a relatively timely manner. The NWS, The Weather Channel, AccuWeather, and several others provide weather alert instant messages. Some simply require that your phone be able to receive text messages while others, such as AccuWeather, require the phone to be internet capable. Along with watch and warning information, local SKYWARN groups may want to consider using text messaging in addition to with other methods such as email and radio, to alert members of a SKYWARN activation. Software packages are available that allow bulk text messages to be sent.

Internet capability is a valuable tool, especially for the mobile spotter. Access to live radar imagery for the storm spotter can be a life saver. This can give mobile spotters an idea of what weather might be heading their way. This is important for the mobile spotter for two reasons. First is safety. By being able to see what is coming, they can determine where to set up safely so they are not hit by an unexpected storm. Second it gives them the information they need to know about where to set up to be able to get the best viewpoint. The thing that the mobile spotter must take into consideration if they are going to use mobile internet is that it adds another distraction inside the vehicle. The mobile spotter needs to keep alert to possibly rapidly changing weather conditions. They must also exercise more caution while driving due to weather conditions. For these reasons it is best to use this only when there are two spotters in the vehicle. One can keep eyes on the road and the other acts as the spotter. You should also follow all applicable laws regarding use of a mobile device while operating a vehicle. Storm spotting does not excuse you from these laws.

There are drawbacks to cellular telephones though. Cellular networks can become overloaded during disasters, preventing voice, data and text communications. Today's smartphones require high bandwidth to support their capabilities and this bandwidth need tends to go up during severe weather as people turn to their phones for information. During a disaster, cellular networks often see spikes in usage that cause interruptions in service. Examples of this increase in usage can be seen during the Minneapolis bridge collapse in 2007 when cellular use was 10 times the average usage[11] and eleven times the average during a school shooting in Ottawa, Canada.[12]

Cellular technology and amateur radio share some commonalities. Cellular service relies on the use of designated frequencies and transceivers to handle the traffic coming in. There is a finite amount of traffic that a cellular tower can handle

A handheld anemometer is a useful tool for the storm spotter. Basic models such as this are readily available at a reasonable price. [Mike Corey, KI1U, photo]

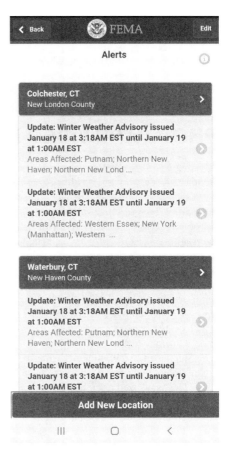

Weather alerts via the FEMA app. [Mike Corey, KI1U, photo]

before it becomes overloaded. And there are other cellular sites in the vicinity that may be sharing frequency spectrum.

The amateur radio storm spotter will typically rely on passing messages via radio. There are times, though, when relaying the message by other means may be necessary. Cellular phones can be valuable tools, but when cellular service goes down the spotter should know that there are still options available. Don't underestimate the value of smartphone technology, but remember it's another tool in the communications tool box.

NWS maintains a listing of third party sources that deliver email and SMS weather alerts to the public.[13]

PHOTO AND VIDEO EQUIPMENT

When we make a report of severe weather, in several ways it is like being the witness to a car accident. The witness is a key part to the understanding of what is happening. He or she provides the investigator with an account of what happened leading up to an accident and what followed. There are factors that have to be taken into account when interviewing a witness. The mental state of the witness must be considered. He or she may be under stress or traumatized from the incident. This may influence the reports on what happened. And frequently witnesses are anxious to talk about what they have seen. This is utilized by police officers as they arrive on the scene. When witnesses are interviewed they are isolated from each other. This allows for an unbiased statement of what was seen and allows the interviewer to compare statements. Often the interviewer can make more progress by communicating to the witness that their assistance may be used to help prevent future accidents. A person is more likely to talk if they feel that their contribution is helpful.

A storm spotter's report is, in many ways, like the witness to the accident. When the storm spotter reports sighting a funnel cloud, it begins a series of events that aid meteorologists in putting together a timeline. One spotter reports a rotating wall cloud. Several minutes later another spotter reports a funnel cloud. A few minutes go by and the second spotter reports a tornado on the ground. These reports are coming in with location, time, and direction of travel. They are being confirmed by other reports and radar data. As the information comes into the NWS a timeline of events can be determined. There are factors, though, that can influence a spotter's report. Like the accident witness, mental state can be a factor. For example let's say the report of a tornado on the ground came from a spotter who was directly in its path. Of course a spotter should take every precaution to not get into this situation, but it can happen.

Being in the path of a tornado will likely affect the mental state of the spotter. And like the witness, spotters are anxious to talk about what they see. That, after all, is why spotters are out there! And again, like the witness, spotters are in essence interviewed in a form of isolation from each other. If I am the only spotter at location X then my report is independent from a report from a spotter at location Y. And in our final comparison with the witness, spotter reports are used to assist efforts to protect life and property. What we report is helpful.

Storm spotters should be aware that there are factors that influence the reports they submit. The most obvious is experience. We can look at pictures of a wall cloud all day long, but Mother Nature has a tendency to throw curve balls and they don't always look like the pictures. As we gain experience as a spotter and with proper training, the accuracy of our reports improves. Stress and anxiety can also play a part in our reporting. When we are under stress our rate of speech naturally increases. This can make communicating effectively more difficult. Our proximity can also influence reporting. Are we looking at a funnel cloud or a wall cloud? If there are trees or structures in the way we may not be able to observe the whole picture. Environmental conditions can also influence reporting. In some parts of the country tornadoes may be rain wrapped, making them difficult to see. But the overall factor in reporting is the individual. We have differing abilities in communicating what we actually see. Some can verbalize what they see with amazing accuracy. 911 operators call this "painting a mental picture" of an event. Others have some difficulty in verbalizing what they see. And, being human, we are all prone to error.

So what can we do to help overcome these limiting factors so that our reports are accurate and more valuable? By showing others what we see, along with a report, we can add accuracy and value to our "witness statement." Here we are in a different situation from the witness. Rarely does an accident allow enough time for a witness to grab a camera. When we go out spotting, though, we have a rough idea of what we're looking for and enough time to bring the tools we need. Let's take a look at how images and video can be used while storm spotting, as well as how we share and preserve those images.

Digital Cameras

For storm spotters a digital camera can be a valuable tool. Most smartphones have excellent cameras built-in, allowing us to share images quickly. Compact digital cameras are also widely available and are quite useful to the spotter. There are some things about digital cameras that a spotter needs to take into consideration.

First let's look at the zoom feature on a digital camera. The ability to zoom in can help make up for distance from an object or allow us to focus in on a particular aspect of an object. Let's say we're looking at what we think is a wall cloud that we are observing from a safe distance. By taking a shot zoomed out we can see the wall cloud in relation to everything else. By zooming in we can get a closer look at the cloud. The quality of the zoomed in picture depends on a couple factors. First is optical zoom versus digital zoom and second is stability. Optical zoom relies on the lens of the camera to zoom in on an object. It is just like zoom lenses on film cameras; elements inside the lens move to change the focal length of the lens. The focal length is then measured in millimeters, so a common zoom lens may range from 18 – 55 mm.

A digital zoom is not really a true zoom lens. Digital zoom basically crops in on a particular area of the image much like photo editing software does. Once it crops in, it enlarges the cropped image to full screen. The trade-off is that some image resolution can be lost doing this. When using a digital camera's zoom feature it is better to utilize optical zoom and save the rest for photo editing software. Zoom is given in a magnification measure such as 10×. The camera will have an optical zoom rating and a digital zoom rating. For example a camera may be rated at 12× optical and 4× digital. This would be preferable to a camera that is 4× optical and 4× digital.

Stability also is a factor when zooming in on an object. As you zoom in, increasing focal length, any movement in the camera becomes exaggerated. Taking the picture at a focal length of 50 mm may not require any stabilization of the camera. As the focal length increases, movement is exaggerated and possibly the amount of light getting into the lens decreases, making it necessary to use a tripod or monopod. Some cameras and lenses feature image stabilization that can help keep the image stable.

To this point the discussion on zoom capability has focused on standard cameras. Cell phones also have zoom capability, but generally are limited to digital zoom. There have been some attempts at adding optical zoom to smartphone cameras with limited success.

A factor that storm spotters may have to take into consideration is low light or night conditions. Trying to take a still image at night can be a serious challenge during good weather, and can be dangerous during severe weather. One way to get useful images at night or in low light is by using the video capture capability that many digital cameras have. The video feature on a digital camera typically does not offer the same quality as a true video camera, but it can still render useful images. During a tornado outbreak a storm spotter used the video mode on a digital camera to capture about three minutes of storm footage. In most of the video all you see is cars passing on the highway and an occasional flash of lightning. But in one brief flash you could make out the silhouette of a tornado. By loading the video into video editing software or using basic desktop tools you could take out a still image of the tornado. The image is not going to be high resolution and may be grainy or fuzzy but it is still useful. In addition to the drawback of low resolution, video mode on digital cameras and smartphones can be limited by storage space.

The next issue we need to look at is memory for the camera. All digital cameras share one feature: they have to have something to store images on. Some have a small amount of built-in memory, but generally they rely on some kind of memory card to store images. As this was written in 2020, most digital cameras use SD cards. SD cards are widely available in capacities ranging up to 512 GB, although cards in the 32 GB to 128 GB range are most economical. As you can see from **Table 3.2**, even at maximum resolution (in MP, megapixels), storage is not an issue.

Now let's look at resolution. The best framework to use to understand resolution is by comparing the final digital image with one taken by a 35 mm film camera. Resolution starts with the camera's rating in megapixels (MP). But since today's cameras, and even many smartphone cameras, are typically rated at 8 MP or higher, resolution is not a concern unless you are printing poster size images. For example a camera with less than 1 MP capability can capture photo quality images for 4 × 6 inch prints. A 5 MP camera can render photo quality prints up to 8 × 10 inches. And at 10 MP, a print 20 × 30 inches can be made at photo quality. Since most of the images a storm spotter is likely to take will be emailed or uploaded to share with the NWS and other spotters, every digital camera on today's market will meet those needs.

Something else for the storm spotter to keep in mind is the effect of rain and precipitation on a camera. Most consumer grade cameras do not like being wet. If water gets on the surface of the lens it can obscure the image. Waterproof and water resistant housings are available for a wide range of cameras. Also it is possible to use a plastic grocery bag, wrapped around the camera body, to make it semi-water resistant.

The final thing to consider with memory is images that will be shared with others. Generally images from a standard camera will need to be transferred to a computer or other device. Many cameras come with a cable that allows direct transfer from the camera to the computer without the need of removing the memory card, and some have built-in Wi-Fi capability. There are also card readers that connect to the computer via a USB port and can read a variety of memory cards. Some computers have this feature built in.

Smartphones

The preceding discussion applies mainly to digital cameras, but today we typically capture images and video with our smartphones. Although some may still keep a dedicated digital camera for weather photography, today's smartphones can capture excellent, high-resolution still images and video in a variety of lighting conditions. They include features such as

View of a tornado at night silhouetted by lightning, captured by the video mode on a digital camera. [Scott Smith, photo]

Table 3.2
Digital Camera Memory Card Capacity for JPG Still Pictures

Photo Resolution	File Size	Memory Card Size/Approx Number of Photos				
		8 GB	16 GB	32 GB	64 GB	128 GB
8 MP	2.4 MB	2,850	5,700	11,400	22,800	45,600
16 MP	4.8 MB	1,425	2,850	5,700	11,400	22,800
22 MP	6.6 MB	1,025	2,050	4,100	8,200	16,400

Hail photographed with a smartphone and shared with NWS via WhatsApp. [Katherine Corey, photo]

zoom capability, image stabilization, and flash photography. Image processing apps can help enhance photos as well.

In addition, smartphones allow us to instantly share images via email, social media, text or even via live video chat. Cloud storage can also be utilized for long term, secure storage. And services like Dropbox allow automatic backup of images from a smartphone camera.

Image Sharing and Storage

The images we capture while storm spotting can do a lot to help back up the reports we submit, and they become part of a well rounded ground-truth report on what the weather is doing. So, let's walk through the steps from capturing an image to sharing it as part of the reporting process, and how and where that image gets preserved.

It's an October afternoon and there have been a few passing showers, but nothing too serious. The SKYWARN net is not active as the threat of severe weather is minimal. A trained spotter notices that it has started to hail and snaps a quick picture; they cannot report it on the net since no activation has occurred. The photo, which at this point is the only report, is shared with the SKYWARN coordinator via the text messaging app WhatsApp. The SKYWARN coordinator then sends it via email to the forecasters at the local National Weather Service. The reported hail gives the forecasters an indication that severe weather may be building in the area. It is the first clue of a severe thunderstorm. In large part due to that one photo, which was sent in along with location and time of occurrence, forecasters were able to get ahead of severe weather and issue appropriate watches and advisories. Sometimes it takes just one report.

The photo was taken with a cell phone's camera and shared with an app that is regularly used for text communication. The SKYWARN coordinator planned for leveraging all forms of communications for reports and could act on one that came in outside of regular activation. The photo was also shared via social media and sent to other spotters in the area directly as a "heads up" for severe weather and possible full activation of the SKYWARN net.

Images captured via cell phones have the distinct advantage of being instantly sharable and storable. Apps such as WhatsApp, Telegram, GroupMe, and Facebook Messenger allow for immediate direct sharing. Combine this with the power of social media such as Twitter, Facebook, and Instagram, and one photo can be amplified to a very broad audience. The key, though, is that an accurate report must still go with the photo. In this case, the additional information needed was size of hail, location, time, and duration.

The photo submitted also became part of several digital archives. The app itself made a copy and stored the image within the phone. Additionally, images can be set to automatically upload to other apps such as Dropbox or Google Photos. And, there are added archival resources available through the SKYWARN coordinator and NWS staff, not to mention that ability of social media to perpetually preserve the image.

While it is easy to think that apps and social media present a challenge to amateur radio storm spotting, the reality is that they are force multipliers for amateur radio, another tool in the toolbox.

OTHER USEFUL EQUIPMENT

Lighting

There are times when storm spotter activation occurs during nighttime hours. Storm spotting at night, especially mobile, presents even more risks to a spotter than during daytime hours. Without a doubt it is far preferable to spot from a fixed location any time visibility is limited. However, if a spotter is mobile there are certain steps they can take to make sure they can see and be seen. And for the home-based spotter or net control station proper lighting is also critical.

First let's look at interior lighting in the mobile environment. Let's consider interior lighting in the car. Interior lighting can come from several sources: installed interior lights in the vehicle, radios, GPS navigation units, or the stereo or dashboard screen. The effect on the driver from this interior lighting can take two forms. First, it can change the driver's ability to adapt visually. Second, the driver can experience what is called "veiling stimulation" where light is reflected off the windshield. In either case there is a potential for the driver's vision to be affected negatively. When storm spotting, mobile, at night we must be aware of how lights on the inside of the vehicle can hinder our vision and ability to react to what is happening outside of the vehicle. We must remember that the driver's first responsibility is safety, not only for those inside the vehicle but also safety on the road.[14]

To help minimize visual distraction to the driver at night, we must minimize interior lighting whenever possible. Mobile transceivers used for storm spotting should have their displays set to a lower light level than normal, but not to the point that they cannot be seen. The same can be said for the car stereo system, climate controls, or dashboard. Many mobile GPS units have a nighttime setting that should be utilized. The vehicle's standard interior lighting should be used only when necessary. One thing that can be used to help minimize the visual effects is a flashlight with a red filter. This can be handy for reading maps and adjusting radios or other equipment.

Another issue for storm spotters in a mobile environment is external vehicle lighting. Of course one thing we must make part of our readiness check list is to make sure all of the vehicle's external lights are in working order, including head and tail lights, brake lights, turn signals, and marker lights.

Some storm spotters include on their vehicle an additional flashing light similar to those found on public safety vehicles or public utility vehicles. The reason for this is so that they can be seen at night by other drivers while stationary, perhaps on a road side, to observe weather. There are several problems with this practice. First, use of these types of lights may be prohibited except by emergency vehicles. While storm spotters do provide a valuable service they do not qualify as emergency vehicles. You should first check with local law enforcement or emergency management before using such lights. Second, a flashing light outside the vehicle can have an effect similar to lights inside the vehicle and negatively affect the driver's ability to see. And third, there is no reason a storm spotter should set up stationary in a place where other drivers may have difficulty seeing them. The storm spotter should look for a place that is safe and out of the way of traffic, such as a parking lot. If they must pull off the side of the road they should keep their headlights on and make use of hazard lights.

Safety is our number one priority when storm spotting. If road and weather conditions make it unsafe to spot mobile at night, stay in a safe location. Our ability to see well at night is very limited and under rapidly changing weather conditions it can quickly become too dangerous to be out on the road. If you do storm spot at night, don't forget to take a storm spotting partner with you. The driver should focus on driving; let the passenger act as spotter and navigator.

The home-based spotter and net control station also have to be aware of lighting issues. Power outages can quickly put us in the dark. Keeping emergency lighting on hand for such occasions is a must. When we look at the design of our home station or net control station we usually look at providing power backup for radios and computers. The power backup may come from a generator, batteries, or a UPS for computers. We should factor emergency lighting into the design. This is usually done for stations that can switch over to generator automatically when the power goes out, such as net control stations operating from an EOC. Not all home-based stations are designed to go to automatic generator backup when the power goes out. Many home stations may have a generator available or have a battery backup system. We can save some time fumbling in the dark by keeping a lighting source readily available for our backup power system.

Binoculars

Another handy tool the storm spotter should keep on hand is a pair of binoculars. We may need to get a closer look at some attribute of a storm or other weather event and binoculars can help. So what binoculars should we use?

To answer this question we first need to understand some binocular terms. Almost all binoculars are described by a pair of numbers such as 7×35 or 10×50. These two numbers tell us two things about the binoculars: magnification and diameter of the objective lens (the lens furthest from the eyepiece).

The magnification tells us how many times an object will be magnified. A 7× means that an object will appear seven times closer than if viewed with the naked eye. There are also zoom binoculars available that have a range of magnification.

The second number refers to the diameter of the objective lens in millimeters. A 10×50 would have an objective lens that is 50 millimeters in diameter. This factor is also important in choosing a pair of binoculars because it tells you how much light can be gathered into the lens. Since we will most likely be using these binoculars during overcast, cloudy, or possibly night time conditions we will want something that will gather as much light as possible. At the minimum we would likely want to use a pair of binoculars with an objective lens of 50 millimeters or more.

Some final considerations on choosing binoculars are cost and size. Binoculars can range in price from well under $100 to hundreds or thousands of dollars. And if we are going to use them while in a vehicle (not while driving) we may consider size when choosing a pair.

The Go-Kit

Most radio amateurs who are involved in emergency and disaster communications are familiar with the concept of a "go-kit." These kits contain essential items used during emergency communications deployments. The kit, which is custom built and stocked by the individual operator, can contain a variety of items for deployments lasting anywhere from a few hours to a few days. The go-kit concept can be valuable to amateurs involved in storm spotting too.

For storm spotters who go mobile when activated, a severe weather go-kit can be a valuable and time-saving tool. What would go into a severe weather go-kit? There are certain tools a mobile storm spotter is likely to use during most activations, such as field binoculars, small camera, severe weather guide card, flashlight, SKYWARN manual, handheld GPS, manuals for radios and other electronics, spare batteries, and bottled water. What you put in your go-kit is entirely up to you.

Also don't forget to keep your vehicle ready for emergencies. A kit with jumper cables, flashlight, first aid kit, fire extinguisher and other such emergency items should always go with you.

This go-kit is capable of HF/VHF/UHF operation and is well suited to operating as net control from almost anywhere. [Frank O'Laughlin, WQ1O, photo]

ONLINE RESOURCES FOR THE STORM SPOTTER

The SKYWARN Website

Before we get into general online resources and software for the storm spotter we will first look at setting up a SKYWARN webpage. Almost all of the NWS WFOs have a page on their website dedicated to SKYWARN and storm spotting. But at the local level a SKYWARN page can also be set up to addresses local issues and provide information on training and meetings. A dedicated web page provides one location where local spotters can find pertinent weather information or links to information, safety guidelines, the local SKYWARN manual and net control guide, as well as any other information that may be useful. And the SKYWARN page can be linked with the local amateur radio club page or ARES/RACES page providing a great recruiting tool for amateur radio and serve as a link between amateur radio storm spotters and non-amateur radio storm spotters.

Other Web Resources

There is no shortage of weather information available on the internet or via smartphone apps. Weather data is available from commercial websites, government sources, schools, media (traditional and social), and from individuals who upload data from home weather stations. Data can be numerical (temperature, pressure, wind speed, and so on), visual (webcams and still images), radar, satellite, and text. Today getting weather information is like drinking from a fire hose!

In this section we will look at some web resources that are available and useful to the storm spotter. Much of the data available through these sites are free, while some require a paid subscription. We will cover a general overview of what is available and how they can be useful to the storm spotter. We will start with the National Weather Service.

NWS Websites

The NWS is the federal agency tasked with providing weather forecasts and warnings for the United States and its territories and waters. The data available through the NWS is about as reliable as you can get. There are several NWS websites storm spotters should be familiar with, starting with the local Weather Forecast Office (WFO). The NWS has 122 WFOs throughout the United States and territories.

The local WFO websites contain a

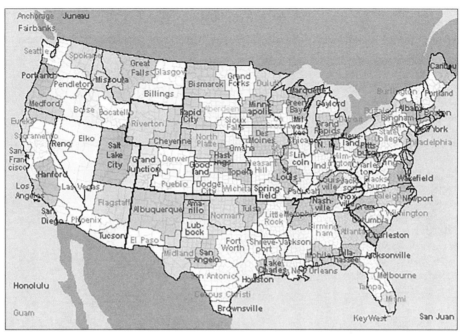

Map of NWS WFOs across the continental US. [National Weather Service]

NWS Central Region home page. [National Weather Service]

wealth of information on their particular county warning areas (CWAs). Typically, a WFO's website will feature meteorological information on current hazards, current conditions, radar imagery, forecasts, and climate. Information on weather safety and education and outreach are also available. Most WFOs also have a page dedicated to SKYWARN, Storm Ready Communities, and general preparedness guides.

Usually a map appears when you log on to the local WFO home page showing the CWA and any watches, warnings, advisories, short term forecasts, or hazardous weather outlooks. By clicking on your county you can bring up a page showing current conditions, the short-term forecast, links to text of short-term forecasts or hazardous weather outlooks, and radar and satellite data links.

Another important feature available on most WFO websites is a link to submit storm reports. This page usually has drop down menus with set terms to describe weather events. This helps prevent any confusion when describing what is going on. For example, "marble size hail" can range in size while "nickel size hail" has an exact size. Along with ability to submit storm reports many WFO websites include information on submitting storm video or photos and provide access to submitted storm reports.

One of the best things a storm spotter can do is to get familiar with the website of their local WFO. It should be a daily practice to log in and check the current forecast and any weather hazards that might be expected.

The next step above the local WFO is the regional office. There are six NWS regional offices: Central, Eastern, Southern, Western, Alaskan, and Pacific. The regional office provides weather information on their particular region of the country. The regional website also serves as portal to national weather information. Regional offices also have available regional publications on weather data and regional weathers programs.

At the national level is the main NWS website.[15] The national website provides information on weather forecasts, hazards, warnings, and observations across the country. There is also information available on NWS sponsored programs such as SKYWARN. The NWS has also created a useful national preparedness web portal known as their Weather-Ready Nation (WRN) initiative.[16] National news and safety campaigns are promoted through

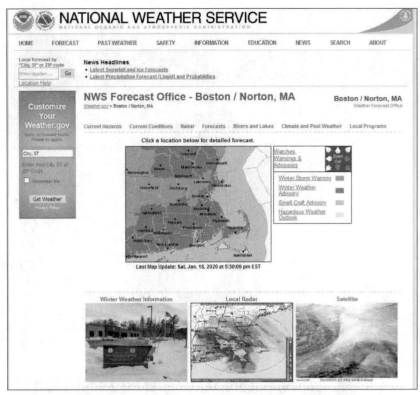

Home page of NWS Boston/Norton WFO. [National Weather Service]

this site. There is also information for organizations such as local SKYWARN groups on becoming Weather-Ready Nation Ambassadors. Several SKYWARN groups have already become Weather-Ready Nation Ambassadors. The NWS sends Ambassadors information relevant to the Weather-Ready Nation strategic priority, such as toolkits for preparedness weeks and planning information for WRN-sponsored events. If you cannot find what you are looking for at your local WFO website then check these national NWS sites.

Specialized NWS Offices

The Storm Prediction Center (SPC) located in Norman, Oklahoma, provides short-term weather forecasts to the public and the NWS WFOs. The SPC is responsible for issuing tornado and severe thunderstorm watches for the lower 48 states and severe weather outlooks. The SPC website[17] provides access to several different products: overview of severe weather across the US, convective outlooks, watches, mesoscale discussions, warnings and advisories, storm reports, mesoscale analysis, and fire weather information. The SPC website also provides access to research and forecast tools.

For the storm spotter, the SPC is a valuable source for seeing what the weather picture is like beyond the local level and what weather hazards may be developing. The storm reports the SPC collects, which come from local WFOs, also provide a great recap of severe weather events. The SPC homepage also provides weather related news articles.

For hurricane and tropical weather there are two NWS internet resources — the National Hurricane Center (NHC) located in Miami, Florida, and the Central Pacific Hurricane Center (CPHC) located in Honolulu, Hawaii. The NHC is responsible for all meteorological decisions concerning forecasting of tropical and subtropical systems for the Atlantic Ocean and the eastern Pacific Ocean north of the equator and east of 140° West longitude. The CPHC is responsible for all meteorological decisions concerning forecasting of tropical and subtropical systems for the Pacific Ocean north of the equator from 140° West to 180° West longitude. The mission of the NHC is to issue watches, warnings, forecasts, and analysis of hazardous tropical weather and to conduct research.[18]

Beyond issuing watches, warnings, and advisories, the NHC and CPHC offer several products on their website valuable to storm spotters in areas affected by

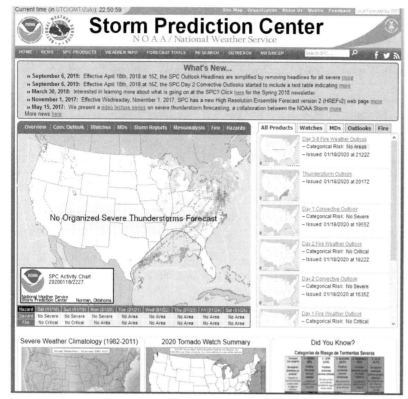

Storm Prediction Center home page. [National Weather Service]

hurricanes. From the main page information can be accessed on current areas of interest for development of severe tropical weather. Satellite and radar imagery is accessible as well as text data. Marine forecasts are also accessible. Information on preparedness and historical data on storms can also be found on the website.

Interactive NWS

Another valuable program available from the NWS is *interactive NWS (iNWS)*. iNWS is intended for members of community emergency planning and response management including SKYWARN net control operators. This website allows registered users to have mobile access to several NWS products.

• **iNWS Alert** sends text message and email severe weather alerts to users. A user may configure it to only send certain alerts, such as tornado warnings and thunderstorm warnings to specific locations.

• **AHPS Mobile (Advanced Hydrologic Prediction Service)** gives users mobile access to hydrographs, current and forecasted stages, and river impacts. "River Watch Points" can also be set up to monitor local points of interest. This feature is handy for areas prone to river flooding.

• **iNWS Mobile** is a Java-based interface that runs on mobile devices and provides access to weather service watches, warning and advisories, radar and satellite imagery, observation, and point forecasts. This feature may not work on all mobile phones. iCWSU is similar to iNWS Mobile but is geared for the aviation community. It provides information on hub forecasts and discussions, TAFs and METARs, storm summaries, convective and tropical outlooks, aviation hazard graphics and imagery.

• **iNWS Mobile Web** is a mobile web application that allows users to access commonly needed NWS products such as weather information by ZIP code or city, current watches, warnings and advisories, current conditions, radar and satellite feeds, and National Weather Service forecasts.

FEMA's Smartphone Application

Another free useful program is the FEMA App.[19] NWS worked with FEMA to ensure all of its watches, warnings and advisories can be received for up to five locations (counties) across the US. The FEMA App also allows the user to upload and share photos of damage and recovery efforts, provides customized emergency safety information, and allows you to locate and receive driving directions to open shelters and disaster recovery centers. The app defaults to Spanish language content for smartphones that have Spanish set as their default language.

WEATHER WEBSITES

There are several commercial websites that provide reliable weather information. They do not require the user to pay any fees, download software, or subscribe to a service, although they may have some features that require a subscription fee, special software, or signing up for an account. We will consider the pros and cons of both free information and subscription services. In this section we will look at a few of these and some of the features that may be useful to a storm spotter.

Almost all of these sites provide some of the same basic information — radar and satellite data, watches and warnings issued by the NWS, local conditions, and local forecast information. Most also offer a mobile service as well. Instead of looking at basic common features, we will look at some of the unique features these websites offer. Of course over time features come and go, so the best way to see what unique features each site offers is to log in and explore. Which site or sites you select for weather information is up to you. Generally most of these services, as well as many more, are available as apps for smartphones and mobile devices.

AccuWeather

AccuWeather was founded in 1962 by Dr Joel Myers and is headquartered in State College, Pennsylvania.[20] Their online service AccuWeather.com provides weather forecasts for the entire United States and over two million worldwide locations.[21] Originally the service provided weather forecasts to utility companies, but in 1971 was expanded to provide weather forecast information to media. Today they provide weather forecast information to media, internet users, newspapers, wireless customers, and government clients.

Through the AccuWeather.com online site, users can access weather products including forecasts, radar, maps, video, news, and extreme weather. Access to these items are common for most weather websites, but there are features that are unique to AccuWeather.

If we look under the forecast features we will find AccuPOP. AccuPOP can provide the probability of precipitation of rain, snow, or ice over a 96 hour forecast period for any location in the United States. This can be a valuable tool for storm spotters. On a day to day basis it can be used for determining the likelihood of precipitation. During periods of heavy rain, ice, or snow it can give us an idea of what to expect over the following hours or days.

Under the radar category there is a useful feature called MapSpace. This combines radar data with a dynamic map similar to maps found on Google Earth. The user can manipulate the maps in several ways — drag to a new location, enter specific location information, change weather data displayed on map, change transparency levels, zoom in and out, and animate the map.

Many key features including AccuPOP and MapSpace are available for no charge. AccuWeather.com also has a paid subscription service, AccuWeather.com Premium, that offers users more information. There is also a professional level paid service, but for our purposes we will only look at the Premium service.

The first noticeable difference between the free service and the paid service is the lack of advertising in the paid service. Many sites, both weather and non-weather, can provide data for free because of the advertising space they sell. By paying a subscription fee users can see only the pertinent data without advertisements. There are features beyond ad-free data for paid users. Paid users can also use Storm Timer and Table, which provides timeline information on severe weather. This can be useful for tracking incoming storms and planning storm spotter activation. Additional features of the paid service include historical data, forecast models, customizable features including a homepage with local weather data, and a planner feature.

WeatherBug

WeatherBug is a little different from most online weather services.[22] While features such as local forecast and severe weather information are available on the WeatherBug website, the strength of WeatherBug is local weather observation. WeatherBug uses data from over 8000 tracking stations and 1000 cameras to provide real-time weather observation data. This data is delivered to users through a downloadable desktop application.

WeatherBug started in 1992 as a way of bringing meteorology into schools. Schools could be networked into a system of weather observation data and were also provided material for in-class curriculum use. The schools were later linked to the broadcast meteorology community through partnerships with local network affiliates. In 2000 the desktop application was launched. This made it possible for a broad range of users to access weather observation data as well as forecasts and weather alerts.

The WeatherBug desktop application provides basic weather data: local forecast, severe weather alerts, radar, and so on. Unlike other services, this is done in a separate application that does not require a web browser. It also allows the user to access local cameras which can provide a vi-

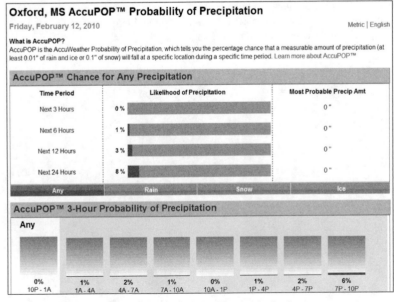

AccuPOP. [Courtesy AccuWeather, used with permission]

sual cue on weather conditions.

Another handy feature that WeatherBug offers is a mobile application for cell phones. Only certain cell phones are supported. The service provides WeatherBug data to mobile users. There are two types of service available: WeatherBug (free) and WeatherBug Elite (paid service). The paid service, like others, offers users an ad-free option.

Stormpulse

Stormpulse was started in 2006 with an initial focus on tropical storm forecasts and information.[23] Since its inception, Stormpulse has evolved to be a highly interactive, easy-to-navigate, graphically sharp weather tool. Its focus has shifted to be a top-of-the-line weather tracking platform for organizations that are impacted by weather.

The data for the site comes from a wide range of reputable sources. Storm track and forecasted path data come from the National Hurricane Center. Cloud cover imagery comes from the NERC Satellite Station at the University of Dundee. Base imagery is provided by NASA. Forecast model data are compiled and retrieved from the South Florida Water Management District.

Storm spotters, particularly those in hurricane prone areas, can benefit from the data provided on Stormpulse in several ways. In real time we can immediately see where the storm is located and a time line showing where the storm is likely to track. Data such as current wind speed, pressure, movement, and category are also provided. On the map we can overlay a variety of data to give us a more robust picture on what the storm is doing and what it might do; forecast models, radar data, and cloud cover can be applied. We can also apply historical tracks, map grid, ocean buoys, wind fields, and map labels. These features allow the storm spotter to follow the storm and make necessary preparations.

The archive data can also be a useful tool for storm spotters. After a hurricane our memory of what actually occurred often becomes a little less than "20/20." And as time moves on exact details may be forgotten. By accessing historical data we can get snapshots of storm data along its path. If we combine this with our logs,

WeatherBug desktop application. [WeatherBug image]

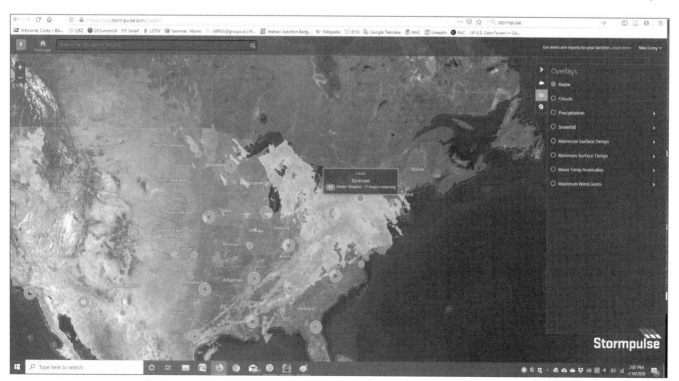

Stormpulse website. [Courtesy Stormpulse, used with permission]

records, and after action reports we can easily build a solid local archive of the storm and how we responded to it. Additionally this gives us the tools to compare one storm to another. We can compare the storm's impact and how we responded to each.

Another useful feature is Stormpulse API. This feature allows hurricane tracking maps to be embedded in a web page. This can be handy for SKYWARN groups in coastal areas prone to hurricanes and tropical storms.

Stormpulse is freely available and offers users an amazing array of features. Like other sites Stormpulse offers paid, ad-free options too. At the Pro level users gain several additional features, including 12-hour satellite loops, auto refresh, tropical watches and warnings, and the ability to save your view.

Weather Underground

Like other free internet weather sites Weather Underground traces its roots to an effort to help bring weather into the classroom.[24] Weather Underground started at the University of Michigan in 1991 as a telnet-based system for real-time weather information. Over time the number of users and information available grew substantially. By 1995 Weather Underground became a separate entity from the university and was providing real-time weather data for 550 US cities. Today Weather Underground provides a wide range of weather information to a wide range of users. The website features access to several standard sources of weather data — local forecasts, radar and satellite data, tropical storm information, information about specific weather events such as hurricanes and tornadoes, and historical information. Weather Underground also features a page where users can access NOAA Weather Radio and listen online.

There are a couple features unique to Weather Underground that are worth noting: local weather stations and the WunderMap. Weather Underground is not alone in providing access to local weather stations. But one handy feature they do provide is a guide on setting up a weather station. The website guides you through choosing a weather station, placing it for accurate measurements, installing and configuring software, and uploading data to the internet. Registered users of Weather Underground can upload data from their weather station and have it accessible online on the Weather Underground website. Weather Underground also shows local weather stations that are not registered with them. You can access data from other local weather stations such as NOAA MADIS (Meteorological Assimilation Data Ingest System) and CWOP (Citizen Weather Observer Program) stations.

WunderMap is a feature that gives the user control over a dynamic map that can access a wide range of weather information. The user can set the map, which is Google Earth based, to view NEXRAD radar, local weather stations, severe weather, webcams, satellite data, United States Geological Survey river data, model data, hurricane, fire, and tornado information, National Digital Forecast Database information, and even photos of the local area. The map also allows the user to select from a standard map, satellite image, hybrid map, or terrain map and has zoom controls.

The Weather Channel

Most people are familiar with The Weather Channel (TWC) and the services that they provide on TV, on mobile devices, and online.[25] TWC's website/app is another great source for weather information for the storm spotter. From their website we can get local forecasts, severe weather information, current weather conditions, alerts, and other information widely available from several reputable sources. But also available online and through television programming is a wide range of educational programs about weather. In the Training chapter we will look at this in more detail.

Weather Nation

Weather Nation is headquartered in Centennial, Colorado, and they pride themselves on 100% of their programming dedicated to weather and forecasting. Weather Nation uses predictive weather modeling sources from NOAA, the Storm Prediction Center, National Hurricane Center, and the National Weather Service. Services are available over via satellite TV providers, through mobile device apps, through streaming services such as Roku, and through smart TVs.

Comparing Sources of Weather Information

There is no shortage of sources freely available online as well as through TV and commercial radio that can provide the storm spotter with weather information. Is one any better than the other? No one can say with absolute certainty and that is definitely not the goal here. Our goal is to get the storm spotter to look at a range of sources. They may find one that they prefer or they may like several. If you're looking for an online source for weather information there are plenty of places to turn to. For day-to-day stuff almost any of these sources, and many others (we didn't even cover local TV station websites), can provide you with the local forecast and current conditions. As storm spotters we have different needs when severe weather threatens. We need up to the minute radar information, timely delivery of watches,

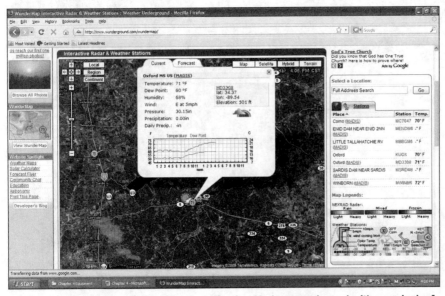

Weather Underground website [Courtesy Weather Underground, used with permission]

warnings, and advisories, and accurate short term forecasts.

Radar imagery from most sources is going to be current. As long as your web browser is refreshed regularly you should have a current image of what is going on. All watches, warnings, and advisories come from the NWS, so no matter where you hear it from it originally came from the NWS. There is almost always a time delay between the issuance of watches, warnings, and advisories and when they appear on the internet. To compensate for this delay it is a better to use NOAA Weather Radio to receive these. There is less lag time when you get it direct from the source. And something to keep in mind with short term forecasts: many online sources deal with weather information at national and international levels. There will be less coverage about what is happening now and in the next few hours through these sites than through local sources such as the local NWS office and local broadcast meteorologists.

SOFTWARE/APPS

In this section we will look at a few programs and apps that are of interest to storm spotters. We will consider functions and features, benefits for the storm spotter, and cost. All of these programs rely on an internet or data connection. This is important because we need to have a connection to servers and other sources that provide the data we're looking for. Some programs also have a communications function that allows us to connect with other spotters and the NWS. All of the programs are standalone programs that operate separate from your web browser or through a smartphone app, although some have both a standalone component and an online, website component. It should be kept in mind that there are many weather applications for smartphones and other mobile devices. Some come from weather companies such as Baron Threat Net, WeatherBug, or AccuWeather, and others are products of local and national media outlets. It is a good idea to try out several to determine one you are comfortable with and make sure it is accurate and useful for your area.

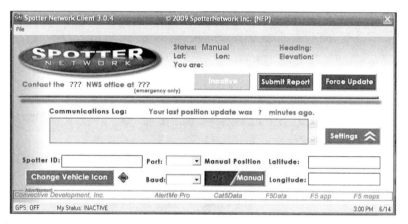

Spotter Network user interface. [Courtesy Spotter Network, used with permission]

The Spotter Network

The Spotter Network (SN), a nonprofit organization not associated with NWS or NOAA, is dedicated to bringing together storm spotters, storm chasers, coordinators and public servants in a seamless network of information.[26] The Spotter Network seeks to provide accurate position data for spotters and chasers for coordination/reporting ground truth to public servants engaged in protection of life and property. The network is a combination of locally installed software for position and status reporting, and web based processing and mapping.

Like technologies such as EchoLink and Skype, the SN also provides a tool storm spotters can use to help fill communication gaps. All that is required to use the SN is an internet connection. Registered users (registration is free) can submit reports online at the SN website or via several smartphone and mobile apps. Reports submitted through the SN are then relayed to the local NWS office. SN also works with a wide range of apps and software programs.

WeatherTAP

WeatherTAP is a commercial weather company that provides, to subscribers, access to a standalone radar tool and a website where additional weather information may be obtained.[27] Subscription to their services does have a monthly cost and users have the option of annual subscriptions. Like most sites there are some unique features to WeatherTAP, such as RadarLab. Radar is one of the most valuable tools storm spotters can utilize. It gives us the ability to see what is coming, what is developing, and where weather threats may be present. Many weather sites and software programs give us base reflectivity and composite reflectivity radar views. These views can be put in motion to give a more complete picture. Additionally we can usually access storm velocity and Vertically Integrated Liquid (VIL).

RadarLab is a standalone Java-based program that gives users access to radar data as well as additional weather information. Users may customize the program for their location and to display certain information on start up.

The radar data available includes base reflective view at several different radar tilts and composite reflective view. Radial velocity and storm velocity data are also available at different tilts. Information on echo tops, precipitation totals, and VIL is also available. The program also allows the user to overlay data on the radar map such as GPS location, place names, roads and highways, and borders.

A handy feature of this program is the

amount of custom data that can be applied to the radar map. Users have the option of including on the map current temperatures, lightning information, watches/warnings/advisories, storm tracks, local storm reports, data from Spotter Network, and surface observations. Beyond the amount of data that can be accessed or applied, the program also allows users to animate, zoom in, change map views and radar sites, and capture still images from the radar screen.

RadarLab HD is also relatively easy and straightforward to use. It does not take long to download, set up, and customize to meet your needs. There are also other versions of RadarLab — Classic and HD with GPS. The website also keeps users updated with news on new features and other information.

Let's take a look at some of the other features that may be useful to storm spotters on WeatherTAP's website.

Users can access forecasts, severe weather information, information for mobile users, and radar and satellite data. Another unique feature is aviation weather information, with access to Terminal Doppler Weather Radar (TDWR). This radar system is used at airports across the country and adds to the number of radar sites users can access. TDWR is a higher resolution radar image (at ranges close in to the radar). Since these sites are at airports near major cities it gives users a higher-resolution image at a city level. There are some advantages and potential disadvantages to using TDWR. A primary function of TDWR is to alert airports of the presence of wind shear, an item of interest to storm spotters. There are also times where NEXRAD radar may be down, and TDWR can be used as a backup so you don't lose local radar imagery. TDWR also is designed to greatly reduce ground clutter. This can be an advantage but it can also reduce views of nearby precipitation that may be of interest.

WeatherTAP also allows users to access aviation forecasts and current conditions. We must remember that the aviation community's interest in weather extends beyond what is happening at the ground level. They also have a keen interest in what is happening in the atmosphere. By looking at upper level conditions we can get a more complete picture of what is going on with the weather.

Another unique feature that is becoming more common among weather websites is KML file access. Users of WeatherTAP can download KML files for Google Earth. These files provide WeatherTAP features in Google Earth.

And, finally, WeatherTAP also provides online guides to the range of products. So if you find yourself stumped between base reflectivity and composite reflectivity you can access an online tutorial to help.

Top 10 Weather Apps

Weather apps for iOS and Android devices are a very popular way to stay informed about the weather and know where severe weather activity is to support SKYWARN Activation when severe weather threatens. Here is a list of Top 10 weather applications for your iPhone or Android phone that a number of SKYWARN spotters and amateur radio operators have given high marks for use doing severe weather. (Any prices shown were current as of mid-2020.)

1Weather

This app is produced by OneLouder Apps and has been around for many years. It is very highly rated in the app stores. It has both daily and hourly forecasts and a plethora of other weather information. The free version is fully functional with ads, but ads can be removed by purchasing the app for $1.99 from the app store.

1Weather app running on an Android device.

RadarScope

RadarScope is a top rated application that can be utilized for detailed interrogation of severe thunderstorms, colorization of wintry precipitation, radar looping as well as severe weather warnings. It provides a fully functional radar software package to your mobile device. To unleash the full power of the application, it has a one time cost of $9.99 to unlock more of the functionality. A Pro Tier 1 subscription plan costs $9.99 per year and includes real-time lightning display, dual pane viewing and longer radar imagery loops.

MyRadar Weather Radar

The MyRadar app is a more simplistic radar app than radarscope but has a full range of features including radar image looping. There are in-app purchases you can make for this app to make it more powerful and they can include things such as a hurricane tracker and additional radar features. This is a good complementary application to other applications that can gather forecasts and severe weather alerts.

MyRadar Weather Radar app running on an iOS device.

RadarScope

RadarScope is a powerful weather radar application that can be utilized on Android, iOS, Apple-TV, and Windows devices. RadarScope has two different tiers of service that can be utilized. Both tiers of service give many features to users with the higher tier of service used for commercial and business purposes. While it's a software application that requires a one-time fee, it is a powerful tool that can be utilized on any device with advanced radar features to interrogate severe thunderstorms to include all the elevation angle of the radars, thunderstorm cell tracks, colorization of wintry precipitation, precipitation maps and many other features. For advanced weather spotters, this app is an excellent software tool that gives both basic and advanced weather radar features that can support proper monitoring of all severe weather hazards.

GRLevelX

GRLevelX from Gibson Ridge offers four programs that are of interest to the storm spotter or anyone interested in weather who wants a high configurable, feature rich standalone program to monitor weather data.[29] Unlike most programs we've discussed, GRLevel2, GRLevel3, GREarth and GR2Analyst are software packages purchased by the user and there are no subscription fees. Let's look at the each program and what it provides.

GRLevel2 is a Windows based program that allows the user to view NEXRAD II

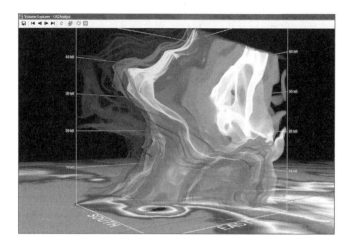

Volumetric display of May 1999 Moore, Oklahoma tornado from GRLevel 2. [Image provided by Michael Laca]

NOAA Weather Unofficial

NOAA Weather Unofficial is a great app to get NOAA Weather forecasts and products, hourly conditions, radar and much more. This is a great way to get all NOAA information in one place via an application. The application includes all severe weather information but it does not have the means to alert the user to severe weather on its own and you do need to manually check for severe weather alerts. Nonetheless, it is a great app to obtain the latest NOAA and National Weather Service information.

Accuweather

The Accuweather application is a strong overall application that has both basic and extended forecasts, hourly forecasts, radar and a MinuteCast feature to predict precipitation on a minute basis. This is an all around weather application that can be used in a variety of situations.

WeatherBug

WeatherBug is an application that has been around for many years. It has all the weather forecast, temperature, radar, weather alerts and much more. It has 18 different weather maps, a lightning alert system and many other features. It also has access to the various weather stations that are WeatherBug weather stations to get real-time weather conditions. The app's user interface is constantly updated to stay current. The pay version of the app is $19.99 but for many weather enthusiasts and spotters, the free version is very feature rich and answers most needs.

The Weather Channel

The Weather Channel app is one of the most popular applications covering all the weather forecast information, current temperatures, future forecasts, severe weather alerts, and radar. It also has breaking weather news features, lightning alerts and much more. The app is available for free, and more features can be purchased for $9.99 for those wanting to use all the app features.

Weather Underground

The Weather Underground has all the basics and it also is one of the premier systems that offers personal weather station information that can be accessed. It also includes weather health information in addition to utilization of the app for severe weather and the personal weather station network. It has a $1.99 per year subscription rate which will eliminate ads when using the apps.

Today Weather

This app was introduced in 2017 and has all of the usual key features such as weather forecasts, humidity, actual and real feel temperature reading, severe weather alerts and expandable weather widgets. It also has an excellent radar user interface. This app is free to download and has optional in-app purchases.

Windy.com

This app is a powerful app that can be used for severe weather is of particular interest to boaters, fishermen and many others with outdoor interests. It can also bring weather models and weather maps of various types for advanced users who want to look at more weather parameters. It also allows access to various webcams for a firsthand look at weather at expected locations. This is another powerful app for weather enthusiasts.

There are many more weather apps currently available and new ones will be developed over time. Mobile weather apps can bring a plethora of information for weather spotters and weather enthusiasts, as well as those just interested in getting an updated weather forecast for personal planning and preparedness.

data. The user can monitor base reflective and velocity, storm relative velocity, and spectrum width sweeps. The software also allows the user to pan, zoom, put radar in motion, and capture screen images. Watch/warning/advisory information is also accessible. The program can also be integrated with a GPS receiver. You can download a fully functional trial of the program from the website, but it is no longer available for purchase and other versions are recommended.

GRLevel3 is similar to GRLevel2 but offers users access to NEXRAD level III radar imagery. It also adds a few extra features in the radar display: composite reflectivity, echo tops, vertically integrated liquid, storm attributes, and rainfall amounts. A fully functional trial can be downloaded from the website and the purchase price as of early 2020 is $79.95.

The next step up is GR2Analyst. At first glance GR2Analyst looks a lot like the GUI for Level 2 and 3, but it offers significantly more features. GR2Analyst offers not only standard two-dimensional radar imagery, but three-dimensional radar modeling of storms. This allows the user to see not only a top-down view of the storm, but a complete picture of the storm's structure. This is called the volumetric display. The volumetric display can give two different views of a storm. The user can take a thunderstorm cell, look at it from just about any angle above ground level, and see the structure of the storm. By adjusting settings, the user may also peel away parts of the storm to virtually see inside. The user may also pull out cross sections of the storm for analysis.

Like the previous two GRLevelX products, the user also has access to a wide range of radar and weather data — base reflectivity and velocity, storm relative velocity, echo tops, vertically integrated liquid, and VIL density. GR2Analyst also gives the user access to maximum expected hail size, probability of severe hail, and normalized rotation. Watch/warning/advisory polygons can also be displayed. The radar data displayed on GR2Analyst comes from NEXRAD level II sites. You can download a trial version of the software and the purchase price as of early 2020 is $250.

The user may add additional customizations to all of the GRLevelX products by adding place files into the program. These place files are available from several sources and can add a variety of new data to the display such as METAR observations, APRS data, SPC watches and outlooks, and local storm reports. Also all of these programs offer GIS support.

GREarth is a Windows viewer for real-time, worldwide weather data with a focus on the continental United States. Features of GREarth include national level 3 radar mosaic, storm tracks, GOES satellite imagery, SPC and NHC products, Spotter Network reports, and high resolution landsat backgrounds just to name a few. The basic program is downloaded from Gibson Ridge and the data stream is purchased through Allison House. The Allison House feed currently costs $180 per year or $15 monthly.

Allison House

A useful add-on for any of the GRLevelX programs or StormLab is a data feed from Allison House. There are several options, at different prices, available: Storm Watcher, Storm Chaser, Storm Analyst, and Storm Hunter. Generally each product offers the same data feeds: surface METARs, offshore surface observations, CWOP (Citizen Weather Observer Program — see below) data, APRS data, GOES satellite images, and so on. The difference is in the types of NEXRAD radar site that each product supports. Storm Watcher offers no data feed from NEXRAD sites, Storm Chaser offers data from NEXRAD level 3 only, Storm Analyst offers data from NEXRAD level 2 and NEXRAD level 2 super-res, and Storm Hunter offers data from all three NEXRAD sites.

To use one of these data feeds, simply select which data you would like to add. You will be given a URL address to copy and paste. Copy the address and paste it into the place file folder in your weather software (GRLevelX or StormLab). This will add that data to the display.

It should be noted that while data is available through subscriptions services such as Allison House, there are also data available for free from a variety of user support groups online.

Citizen Weather Observer Program (CWOP)

CWOP, the Citizen Weather Observer Program, is a private-public partnership between individual citizens who collect weather data on the one hand, and the weather services and homeland security on the other.[30] The program has three objectives: 1) to collect weather data contributed by citizens; 2) to make these data available for weather services and homeland security; and 3) to provide feedback to the data contributors so that they have the tools to check and improve their data quality. The users of CWOP data range from other citizens to universities to government agencies to commercial weather companies. There are over 8000 registered CWOP members around the world. The data is sent via the internet to FindU.com and from there to the NOAA MADIS (Meteorological Assimilation Data Ingest System). To access data from any CWOP

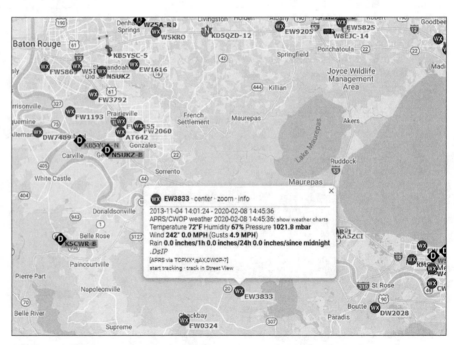

CWOP and APRS data displayed on GRLevel 2. [Used with permission]

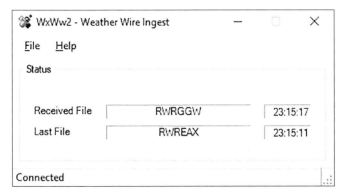

Weather Message Ingest portal (left) and Alert message received (right). [Images provided by Jim Palmer, KB1KQW]

station, all that is required is an internet connection to a site that displays the data such as Weather Underground or weather software such as GRLevelX.

The data from CWOP stations can be a valuable tool for local storm spotters. By checking the data coming from your local CWOP station you can see temperature, pressure, and precipitation. Graphs allow us to also see trends over the last several hours or days. All of this data is stored online so we can also go back and retrieve past data.

The CWOP station consists of four components. First is a weather station that is properly set up and calibrated. Second is a computer that can connect to the weather station. Most weather station manufacturers have cables or a Wi-Fi or Bluetooth interface available for this. Third is software that can gather the data from the station and configure it into a data packet to send to the server. Software is available from weather station manufacturers as well as independent sources. And finally, an internet connection is needed so the data can be uploaded.

As data is uploaded to the internet it is checked for accuracy. You can sign up to receive daily reports on any errors coming from your CWOP station.

Weather Message

Weather Message is a commercially available software package that is comprised of a client-server based system to receive and process all NWS watches and warnings (**www.weathermessage.com**). For a minimal cost, a user can enable a system to receive desktop notifications, email alerts, social media integration and more. Various packages are available to the user depending on the size of the group and their specific needs.

The system administrator begins by defining a list of products that will trigger an alarm once received; the list is placed in the setup file. Additional parameters include defining a client ID, specific states and counties for the warning, alarm settings, and methods of alerting (client, email, fax, and more). By defining alarm settings, the user can trigger customized alerting sounds or actions, including automatic printing of the warning if desired.

Internet data is usually received through a NOAA Weather Wire feed, which can be set up at no cost through the NWS. Previous EMWIN data feeds have been discontinued and are no longer available making Weather Wire the primary method of receiving internet data. Additional methods of receiving data include through NOAAPort and IPAWS. For a method of receiving data from the GOES satellites without relying on public internet, some have even downloaded plans from the internet for using XRIT decoder software, a grid antenna, and SDR receiver to receive and decode the products for ingestion into Weather Message.

The end user can then connect to the server from the client and will receive any products assigned to that client ID. More than one client can connect to the same ID which helps regional groups focus only on watches and warnings that pertain to their area of interest. If an alert is missed, the user is able to request the product from the server, and can request any product the server has received regardless of their client ID.

The paging features of Weather Message range from more traditional pager software to website and social media integration with Facebook and Twitter. This has allowed some weather spotter groups to rapidly alert their members of impending severe weather for a rapid response. In addition, groups involved with emergency management can offer another method of quickly alerting the public.

HOME AND PORTABLE WEATHER STATIONS

For most radio amateurs involved in SKYWARN and storm spotting, our interest in weather goes beyond just severe weather and SKYWARN activations. We generally check the forecast each day. We like to know what is happening now, even when it is clear, warm, and sunny. Our family, friends, and colleagues tend to turn to us to find out what the weather will be like today. We are good storm spotters because we stay informed. We stay informed so the weather doesn't take us by surprise. We stay informed so we are ready to go when called on.

A valuable tool for the storm spotter is a home weather station. Today home weather stations come in all varieties from the simple barometer/thermometer that hangs on a wall in our home or out in the garden

Technician collecting weather data for the US Department of Agriculture. [Scott Bauer, US Department of Agriculture, photo]

Best Weather Stations — Factors to Consider When Making Your Purchase

(reprinted with permission of WeatherShack.com)

• **Weather Variables**. This is the most important factor when purchasing commercial weather stations because the type of weather measurements (variables) you wish to monitor will determine if you should be looking for a basic personal weather station (temperature, humidity, barometric pressure) or a complete professional weather station (wind, rain, and more). You should also consider how a unit's indoor console displays this data.

• **Cost**. How much you are willing or can afford to pay for your weather station is a factor that will impact all of the factors that remain. In general, the more money you can invest in a personal weather station the better it is. However, comparing specifications is definitely worthwhile because the old adage "you get what you pay for" is not necessarily true in all cases.

• **Installation Issues**. The site where you intend to install your personal weather station needs to be evaluated, taking into consideration the distance from the indoor display console to where the sensors have to be located. For cabled weather stations, the length of the cable that comes with the unit will determine the maximum distance. Some manufacturers do offer extension cables for their cabled weather stations. Wireless home weather stations have a maximum "unobstructed" or "line of sight" range rating, which is diminished by the type and number of building materials the signal must penetrate. A rule-of-thumb for a typical installation is one half to one third the distance of the unobstructed range rating. Possible sources of interference should also be taken into consideration with the installation of wireless weather stations. The altitude where your weather station is going to be installed can also be an issue, and you should be aware that many weather stations are limited to operating at 6000 feet or below with regard to accurate barometer readings.

• **Accuracy, Resolution and Range**. These three factors are of primary importance when selecting the best weather stations. Accuracy is how close the displayed measurement reading is to the true measurement value (for example, ±1°F for temperature). Resolution is the smallest increment that the unit is capable of measuring and displaying (for example, 0.1 °F for temperature). Range is the minimum and maximum limits the unit is capable of measuring within (for example, −40 °F to +150 °F for temperature).

• **Update Interval**. This is the rate at which the personal weather station updates the display. Update intervals can vary significantly between units, from as often at once per second to as little as once every three minutes or even longer. Watching the wind data change every three minutes during a variably windy day, or likewise the rain data during a heavy rainfall, could limit the enjoyment you receive from your weather station.

• **Weather Forecasting**. If available, many commercial weather stations base their forecast on the barometric pressure trend (rising or falling), which is not as accurate as taking other variables into account. The more sophisticated professional weather stations takes into account not only barometric pressure, but also wind, rainfall, temperature humidity, and even the latitude and longitude of the station, resulting in a much more accurate forecast.

• **Historical Data**. Most basic weather stations and the less expensive complete weather stations display current data and very little in the way of historical data, perhaps the high and low readings for a period of time (often between manual resets). Weather stations that feature barometric pressure often include a graphic display of the trend for the past 24 hours. More extensive historical data retention is usually only available in a high-end professional weather station. For example, the Davis Vantage Pro2 series can display the highs and lows (and/or totals or averages) for nearly every weather variable for the last 24 hours, days, months, or years!

• **Computer Interface**. If you are interested in connecting your weather station to a computer you will most likely need to purchase a complete weather station, and not all of those come with that capability or option. With computer weather equipment (interface and software), you can record and graphically display weather variables at an interval that's typically user selectable. Depending on the software you can make forecasts, graph weather trends, post data on the Internet, or even send e-mail alerts. Some manufacturers include a "data logger', which has a built-in memory to store weather data for later retrieval. This allows you the flexibility of not having your computer on software running at all times.

• **Weather Station Review**. There is a lot of information available on the Internet about specific models of weather stations. We suggest you perform a search for some reviews on the models you're considering.

to weather stations that are digital and provide information on almost all facets of the current weather condition. The current selection can be quite daunting if you have never purchased a home weather station or weather instruments for home use. A commercial vendor of weather equipment has an excellent guide to choosing a weather station (see the accompanying sidebar).

REFERENCES

[1] NOAA Weather Radio (NWR): **www.weather.gov/media/nwr/NWR_Brochure_NOAA_PA_94062.pdf**
[2] Bob Bruninga, WB4APR: **www.aprs.org**
[3] P. Dodd, G3LDO, *Amateur Radio Mobile Handbook*, 2nd edition (RSGB, 2011), available from **www.arrl.org/shop**.
[4] S. Ford, WB8IMY, *Radios to Go! Getting the Most from Your Handheld Transceiver* (ARRL, 2012).
[5] EchoLink: **www.echolink.org**
[6] J. Taylor, K1RFD, *VoIP: Internet Linking for Radio Amateurs*, 2nd edition (ARRL, 2009).
[7] Google Earth website: **earth.google.com**
[8] Department of Homeland Security: **www.ready.gov/alerts**
[9] Weather-Ready Nation website: **www.weather.gov/wrn/**
[10] FEMA website: **www.fema.gov/integrated-public-alert-warning-system**
[11] T. Wailgum, "Minneapolis Bridge Collapse: Why Cellular Service Goes Down During Disasters," *CIO*, August 3, 2007.
[12] D. Crowe, "Overload!," *Cellular Networking Perspectives*, Q4 2006.
[13] NWS website: **www.weather.gov/enterprise/**
[14] J. Devonshire, M. Flannagan, "Effects of Automotive Interior Lighting on Driver Vision," University of Michigan, Transportation Research Institute, March 2007.
[15] NWS website: **www.weather.gov**
[16] Weather-Ready Nation website: **www.weather.gov/wrn/**
[17] Storm Prediction Center (SPC) website: **www.spc.noaa.gov**
[18] NWS website: **www.nws.noaa.gov/directives/sym/pd01006001curr.pdf**
[19] FEMA website: www.fema.gov/mobile-app
[20] AccuWeather: **www.accuweather.com**
[21] Figures given by AccuWeather.
[22] WeatherBug: **www.weatherbug.com**
[23] Stormpulse: **www.stormpulse.com**
[24] Weather Underground: **www.wunderground.com**
[25] The Weather Channel: **www.weather.com**
[26] Spotter Network: **www.spotternetwork.org**
[27] WeatherTAP: **www.weathertap.com**
[28] WXWarn download: **www.wxspots.com**
[29] GRLevelX: **www.grlevelx.com**
[30] Citizen Weather Observer Program: **www.wxqa.com**

Training

Most storm spotters, whether or not they are amateur radio operators, have taken a SKYWARN class. Some have gone farther with Advanced SKYWARN, when it is offered. A few have taken both SKYWARN classes, participated in exercises, completed ICS training, and have other qualifications. One area where amateur radio operators differ from other storm spotters is with training. It is rare to find a radio amateur who is *just* a storm spotter. Many hams active in storm spotting are also active in ARES, their local Community Emergency Response Team (CERT team), or serve as volunteers with emergency management, Red Cross, or a public safety agency. Because of our involvement in public service, training matters. As storm spotters, the more training we can get, the better prepared we will be for an emergency.

In this section we thoroughly look at training for the amateur radio storm spotter, going beyond the basic level to better prepare ourselves. We divide training up into three basic areas: foundation, readiness, and continuing education. Much of this may overlap with training needed for other public service functions such as ARES or CERT.

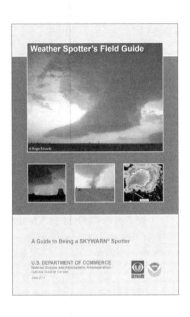

The NWS's *Weather Spotter's Field Guide* offers a general overview of the program. It may be downloaded from www.weather.gov/media/bis/Weather_Spotter_Field_Guide.pdf

SKYWARN FOUNDATION TRAINING

Without a doubt, the first training any storm spotter should receive is through the NWS SKYWARN program. Repeating SKYWARN history from Chapter 1 is unnecessary, but it is worth looking at some current information about the program. As of 2020, NWS SKYWARN has trained more than 350,000 spotters across the United States to provide information to NWS offices about local storms. This network of trained spotters, combined with insight from radar and other resources, aids timely and accurate NWS watches and warnings.

In addition to amateur radio operators, many other groups have a vested interest in severe weather and need accurate and timely information about it. Groups such as police, fire fighters, emergency medical services, public safety dispatchers, media, school officials, hospitals, places of worship, nursing homes, and concerned citizens all make up the corps of SKYWARN volunteers.

Local NWS offices provide free SKYWARN training administered at each office by the Warning Coordination Meteorologist (WCM). The training usually takes only a couple of hours, covering basic information about severe weather: thunderstorm development, identification and structure; identifying storm features; report methods and standards; and basic severe weather safety. The WCM also may customize training to meet local severe weather readiness needs; for example, nor'easters along the North Atlantic coastal area wouldn't be relevant in Iowa.

Many NWS field offices also offer an advanced SKYWARN class that, like the basic course, can be structured to meet local storm spotter needs. It may take a few hours or be spread over two or three days. The advanced course builds on the basic one, providing more details on different types of severe weather, safety, what to report, radar interpretation, and spotting resources. The precise course curriculum is determined locally. Check with your NWS office to see if the advanced course is offered and what it includes. The NWS's *Weather Spotter's Field Guide* offers a general overview of the program.

In some areas there is a need for additional training because of how and to whom reports are sent. For example, spotters may report to net control, who then relays to the local emergency management agency and the National Weather Service. Spotters may also need training on interaction of local agencies during severe weather, local activation procedures, and net protocols. Some in-house training, on top of the NWS SKYWARN program, can help your SKYWARN group respond effectively to severe weather.

THE ARRL EMERGENCY COMMUNICATIONS COURSE

In 2001 the ARRL began an online continuing education program for amateur radio operators in emergency communications. The Amateur Radio Emergency Communications Course (ARECC) was later broken into three different levels to cover the amount and diversity of information available about the subject. The Level 1 course, still available as Introduction to Emergency Communications, covers a wide range of basic topics important to amateur radio emergency communications: served agencies, network theory, communication skills, message handling, nets, operations and logistics, and so on. Whether you are part of an ARES group or a SKYWARN storm spotter, this is essential training. As discussed elsewhere, SKYWARN storm spotters are often involved in their local ARES group.

The EC-001 course is designed to provide basic knowledge and tools for any emergency communications volunteer.

The course has 6 sections with 28 lesson topics. It includes required student activities, a 35-question final assessment and is expected to take approximately 45 hours to complete over a 9-week period. The course is hosted on Canvas, and the student has access to it throughout the nine week window. You must pace yourself to be sure you complete all the required material in the allotted time. There is no cost for this course.

READINESS TRAINING

Emergency Management Courses

A valuable training resource for the storm spotter and emergency communications is the Federal Emergency Management Agency (FEMA). Via the Emergency Management Institute, (**training.fema.gov**), FEMA offers free online training courses ranging from basic level and citizen awareness to advanced responders. Many of these courses are great training resources for storm spotters.

In addition to FEMA, many state and local emergency management agencies offer training classes. Examples include Community Emergency Response Team (CERT), Weather for Emergency Management, National Incident Management System (NIMS)/Incident Command System (ICS), Hurricane Readiness for Inland and Coastal Communities, floods, and disaster recovery. In some areas you must be associated with a public safety or emergency management agency to take these types of courses. To find out course requirements and costs contact your local or state emergency management or homeland security office.

Over the last decade or so ICS/NIMS increasingly has become mandatory training for many emergency responders. In 1970 a wildfire in California raged for 13 days, killed 16 people, destroyed 700 structures, and cost $18 million per day. Several entities involved in the fire response assembled to form a better and more efficient concept for fire response: the Incident Command System. Over time, the use of ICS spread to cover a wide range of incidents, including floods, earthquakes, hazardous materials, and aircraft crashes. For several decades there have been some attempts at blending ICS and other incident management systems into a national standard.

FEMA offers a number of emergency management courses.

In 2003, President Bush issued the Homeland Security Presidential Directive 5 (HSPD-5) which ordered the development of the National Incident Management System (NIMS), taking ICS a step further. The purpose of NIMS is to provide "a consistent nationwide template to enable federal, state, tribal, and local governments, non-governmental organizations (NGOs), and the private sector to work together to prevent, protect against, respond to, recover from, and mitigate the effects of incidents, regardless of cause, size, location, or complexity. This consistency provides the foundation for utilization of NIMS for all incidents, ranging from daily occurrences to incidents requiring a coordinated federal response." NIMS and ICS training are available through FEMA online courses.

Community Emergency Response Team (CERT) was developed by the Los Angeles Fire Department in 1985, under the idea that during the early stages of a disaster, citizens may be on their own. If some in the general public had training in basic disaster survival and rescue skills they would be more likely to survive until responders could arrive. With the assistance of LAFD, FEMA's Emergency Management Institute expanded the program nationwide in 1994 to include all disasters. In 2003 President Bush initiated the Citizen Corps as a way for all Americans to become involved in homeland security efforts. CERT was a perfect vehicle for this program.

CERT training covers basic disaster preparedness and response skills. Disaster related training includes preparedness, fire suppression, medical operations, light search and rescue, psychology and team organization, and a simulation. Through the Citizen Corps website **www.ready.gov/citizen-corps** you can find resources that include what is needed to start, train, and maintain a CERT.

CPR / First Aid / AED

Medical emergencies can arise whenever we are active as storm spotters. A storm spotter could encounter someone struck by lightning, a vehicle accident, injury due to hail or, in large scale disasters, mass injury situations. Storm spotting does not qualify us to render medical aid, but some basic, valuable training can prepare us to deal with medical emergencies until responders can arrive. The Red Cross, American Heart Association, hospitals,

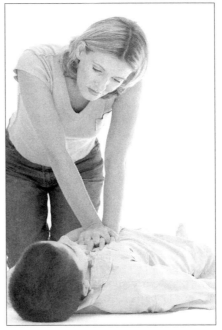

With luck you'll never need to use it, but CPR training is always a good idea.

and fire departments offer classes in most areas, including cardiopulmonary resuscitation (CPR) and basic first aid.

CPR is a combination of chest compressions and artificial respirations into the patient to keep blood flowing through the heart and oxygen into the blood. It is not likely to restart the heart, but keeps oxygenated blood flowing to prevent tissue death and buy time until the heart can be restarted.[1]

Most basic first aid courses cover simple techniques that can be used to stabilize a patient until responders arrive, including recognizing an emergency situation, caring for an unconscious victim, controlling external bleeding, sling and binder skills, removing contaminated gloves, and basic prevention of disease transmission. You should check with your local Red Cross or other basic first aid training provider for more details.

Automated External Defibrillators (AEDs), located in most public places, are used when a victim's heart has stopped. An AED administers a shock to aid in restarting the heart. Using a combination of lights, voice prompts, and text messages, it tells the person administering the AED when to shock. While it is not likely that a mobile storm spotter will be in a location where an AED is present, this is still useful knowledge. AED training often is combined with first aid and CPR training.

CONTINUING EDUCATION

Keeping up to date and expanding our knowledge base helps us to stay effective and prepared. In this section, we will look at some of the abundant resources available to storm spotters in order to learn more about severe weather.

NWS Resources

A key role of the NWS is education.[2] Beyond SKYWARN training, the NWS offers a wide range of educational resources. Some are geared for children; others focus on safety and awareness; and some are quite useful to those interested in general meteorology and severe weather. Let's take a look at some NWS educational tools that a storm spotter may find helpful.

JetStream — Online School for Weather is a website that offers material for educators, emergency managers, or anyone interested in weather and weather safety.[3] The online curriculum covers the atmosphere, the ocean, synoptic meteorology, specific weather events such as tornadoes

The JetStream — Online School for Weather website. [courtesy National Weather Service]

and hurricanes, radar, remote sensing, and weather on the web. The online course features an introduction to each subject, an overview with visual aids, a frequently asked questions section, and review questions. You can go from one topic to the next in order or study whichever one interests you. The course is free and available to anyone.

The MetEd professional development series is a free online weather training from the COMET Program, which is part of the University Corporation for Atmospheric Research (UCAR). MetEd, which receives funding and support from government agencies in North America, Australia, and Europe, is far more in-depth than JetStream or SKYWARN classes. MetEd is intended for operational forecasters, the academic community, and "anyone interested in learning more deeply about meteorology and weather forecasting topics." Such topics include climatology, fire weather, radar meteorology, and space weather. You will need to sign up for a MetEd account to get started.

The NWS training division also offers some online educational resources covering a wide range of topics such as hydrology, Advanced Weather Interactive Processing System (AWIPS), and general meteorology. Most of this material is designed for practicing meteorologists, but storm spotters and others interested in weather may find the topics useful and educational. The NWS Climate Prediction Center also provides educational resources on La Nina and El Nino, climate fact sheets and monographs, and monsoons.

NWS also offers a useful quarterly publication called *Aware*, designed to "enhance communication within the Agency and with the emergency management community." Current and past issues are available free through the NWS website, or you can sign up to have new editions e-mailed to you upon publication. Many local NWS offices also publish newsletters containing information of interest to storm spotters in their County Warning Area (CWA). NWS regional offices are another source for news and information for storm spotters.

Beyond online thematic meteorology training there are other products available from the NWS that are useful for storm spotters, including a wide range of brochures on different weather topics. These brochures provide basic information as well as useful safety guidelines. Throughout the year, the NWS promotes awareness weeks for different types of weather hazards. NWS educational resources are available through the main NWS website and through your local NWS field office.

Colleges and Universities

Throughout the United States many colleges and universities have meteorology programs and research centers. While the primary function of these programs is for academic training and research, many have resources available to the public about meteorology and severe weather.

The University of Illinois Weather World 2010 project (WW2010) is a free, online instructional guide covering meteorology, remote sensing, and reading and interpreting weather maps. The site also hosts open-ended projects, classroom activities and teacher's aids, each providing instruction on a wide range of related topics. For example, the meteorology guide contains sections named Light and Optics, Clouds and Precipitation, Forces and Winds, Air Masses and Fronts, Weather Forecasting, Severe Storms, Hurricanes, El Nino, and Hydrologic Cycle. Each topic contains text, images, video, and other dynamic presentations. The site is reviewed and edited by members of the Department of Atmospheric Sciences at the University of Illinois-Champagne-Urbana and the Illinois State Water Survey.

The WW2010 project is one of many educational outreach programs available through universities and colleges. Most schools with a meteorology or climatology program have material available through their websites that covers different weather topics. Information on many of these programs is available through the University Corporation for Atmospheric Research and the American Meteorological Society websites.

National Hurricane Conference

Held in the spring each year, the National Hurricane Conference offers a wide range of training opportunities for those involved with hurricane response.[4] The primary goal of the National Hurricane Conference is to improve hurricane preparedness, response, recovery and mitigation in order to save lives and property in the United States and the tropical islands of the Caribbean and Pacific. In addition, the conference serves as a national forum for federal, state and local officials to exchange ideas and recommend new policies to improve Emergency Management.

To accomplish these goals, the annual conference emphasizes:
• Lessons learned from hurricane strikes
• State of the art programs worthy of emulation
• New ideas being tested or considered
• Information about new or ongoing assistance programs
• The ABCs of hurricane preparedness, response, recovery and mitigation — in recognition of the fact that there is a continual turnover of emergency management leadership and staff

Held in conjunction with the National Hurricane Conference is a symposium on the amateur radio response to hurricanes. Presenters include ARRL, WX4NHC,

National Hurricane Center director Dr. Rick Knabb presents on the value of amateur radio to the mission of NHC. [Julio Ripoll, WD4R, photo]

Hurricane Watch Net, VOIP WX Net, the Canadian Hurricane Center, and the National Hurricane Center. The symposium is usually live streamed for those who cannot make it to the conference location. Following the presentations there is an opportunity for panel discussion and Q&A.

Journals, Books and Media

Formal, scientific journals are another useful tool to expand your meteorology knowledge, keeping in mind though that most journal articles are written for an academic and professional audience. The publications will be rich in scientific, technical, and professional jargon, but you can glean a great amount of knowledge from these resources. You are not likely to find them at your local library or bookstore, however. Journal subscriptions are usually quite expensive, although some online journals are free, such as *The Electronic Journal of Severe Storms Meteorology*. Most college and university libraries subscribe to journals or electronic resources that can retrieve journal articles. However, most academic libraries require patrons to be faculty, staff, or students to access this information.

The American Meteorological Society is a source for scholarly material if you are not affiliated with a college or university. Founded in 1919, the AMS promotes "the development and dissemination of information and education on the atmospheric and related oceanic and hydrologic sciences and the advancement of their professional applications." Membership is available in several different classes: full, associate, or student. They also publish nine different weather-related journals in print and online and host conferences throughout the year, as well as offer a wide range of products and services to their membership (currently about 14,000).

There are also several books that are of interest to storm spotters looking to expand their meteorological understanding. In 2009, the AMS published *The AMS Weatherbook: The Ultimate Guide to America's Weather*. Written by Jack Williams, a former editor of the *USA Today* weather page, the book covers a wide range of meteorological and climate information such as daily weather patterns, air pollution, and global warming.

The inquisitive storm spotter may also want to pick up a basic meteorology textbook, such as those used for freshman meteorology students. One that has been widely used is *Meteorology Today* by C. David Ahrens. This textbook, which as of 2016 is in its 10th edition, is easily available through online sources. Another textbook and AMS publication, *Weather Studies: An Introduction to the Atmospheric Sciences* by Joseph Moran, may also be useful to storm spotters. Prices for these textbooks are higher than for mass marketed books, but used copies abound.

There are also books dealing with specific weather events and historical weather events that are of interest to storm spotters. Author Peter Chaston has written several meteorology books. Chaston has written books on hurricanes, reading and interpreting weather maps, and thunderstorms, tornadoes, and hail. Chaston also has a book entitled *Weather Basics*. And we cannot forget historical weather events. Peter Felknor wrote about one of the deadliest tornado outbreaks in US history in *The Tri-State Tornado: The Story of America's Greatest Tornado Disaster*. This event, which impacted Missouri, Illinois, and Indiana in March 1925, set the record for the deadliest tornado in US history and the longest tornado in US history (a 219 mile path). A single tornado, which measured an F5 with 200+ MPH winds, was responsible for 234 deaths, 2027 injuries, and the destruction of 15,000 homes.

There is a wide range of meteorology books available. Some are written for a general readership while others are written with meteorologists or meteorology students in mind. Most can be found in public or university libraries or online. There are also many books dealing with weather history, both generally and specific weather events.

Magazines are another good source for meteorological reading. *Weatherwise* magazine is one such meteorology publication. Published quarterly, *Weatherwise* covers topics related to meteorology and climatology. They also publish photo contests, with photos submitted by readers, and an annual review of the previous year's weather. Another magazine source is *BAMS*, the *Bulletin of the American Meteorological Society*, the membership magazine of the AMS. Articles in *BAMS* are written primarily from a scientific perspective by members. Both *Weatherwise* and *BAMS* are often available through college and university libraries or through subscription or membership.

Another source of further reading is government reports on major natural disasters. Reports published by the US government, such as the Congressional report on the response to hurricane Katrina, are freely available online.

And for those who are interested in weather beyond the national level there are sources available through the World Meteorological Organization (WMO). The WMO, like the International Telecommunication Union (ITU), is an organization within the United Nations (UN). The WMO is "the UN system's authoritative voice on the state and behavior of the Earth's atmosphere, its interaction with the oceans, the climate it produces and the resulting distribution of water resources." The WMO was founded in 1950 but traces its roots to the International Meteorological Organization which was founded in 1873. Through the WMO website you can access publications relating to their primary areas of interest: meteorology, climate, oceanography, and water distribution.

Another source of training and preparedness material that cannot be overlooked is video resources. Companies such as The Weather Channel offer a wide range of videos on their websites that are freely available. Topics covered include safety and preparedness, forecasts, and news. Through an online store, they offer videos of programs that air on The Weather Channel such as *Storm Stories* and *When Weather Changed History*. Of course these programs may also be viewed by tuning in to a scheduled broadcast. National Geographic also offers a series of videos for sale that deal with weather: *Forces of Nature, Restless Earth* and *Tornado Intercept*. Also searching Internet sources can bring up a wealth of weather videos — some good, some bad. While on the subject of videos, don't forget websites of storm chasers. Many times they post online video clips from their chases.

Videos can be a great resource, but keep in mind that while some are quite instructive there are some out there that are not. Some may simply show severe weather as it is happening. Others may show people doing dangerous things. Always remember that safety is your number one priority. Unfortunately many of these amateur videos don't come with the warning "don't try this at home." When you're watching such videos you can easily turn them into training opportunities. You can assess the video on safety issues, reportable weather events, or look for specific weather phenomena.

SETS, EXERCISES, AND DRILLS

SKYWARN participants, like ARES members, can benefit by conducting regular exercises and drills and taking part in the annual ARRL Simulated Emergency Test (SET). Each year the ARRL Simulated Emergency Test provides an opportunity for amateur radio operators to test their emergency communications skills. Generally SET is planned for a weekend in the fall, but some areas may plan to conduct a SET at a different time to meet local needs. For example along the Gulf Coast the fall is hurricane season so SETs are generally held in the spring. The planning of SET exercises is generally coordinated by section leadership: Section Manager (SM), Section Emergency Coordinator (SEC), District Emergency Coordinator (DEC), and local Emergency Coordinator (EC). They also play a role in the review of test results.

The *Amateur Radio Public Service Handbook*[5] lists the purposes of the SET as to:

• Help amateurs gain experience in communicating, using standard procedures under simulated emergency conditions, and to experiment with some new concepts.

• Determine strong points, capabilities and limitations in providing emergency communications to improve the response to a real emergency.

• Provide a demonstration, to served agencies and the public through the news media, of the value of amateur radio, particularly in time of need.

And the goals of the SET are to:

• Strengthen VHF-to-HF links at the local level, ensuring the ARES and NTS work in concert.

• Encourage greater use of digital modes for handling high volume traffic and point-to-point welfare messages of the affected simulated-disaster area.

• Implement the Memoranda of Understanding between the ARRL, the users and cooperative agencies.

• Focus energies on ARES communications at the local level. Increase use and recognition of tactical communication on behalf of served agencies; use less amateur-to-amateur formal radiogram traffic.

A SET can be very beneficial to the SKYWARN storm spotter. A SET exercise may be based on a simulated weather emergency. SETs often focus on events such as tornadoes, hurricanes, and other natural disasters. The SET can also help local SKYWARN spotters to fine tune and improve their communications abilities for responding to severe weather. Spotters can learn to relay reports more accurately, and net control and relay stations can identify ways to improve communications with the NWS. And it gives amateur radio operators a good chance to demonstrate their role during severe weather.

The SET also provides an ideal environment to work with served agencies such as the NWS and emergency management. This makes it possible for local amateurs and the served agencies to fine tune and develop their working relationship. Ultimately this can improve communications during a real SKYWARN activation.

The SET is also an ideal opportunity to try out and test new modes and equipment. We can try out local APRS networks, the EchoLink connections with the local NWS office, the D-STAR repeater system, or packet networks and test strengths and limitations. For mobile spotters and home-based spotters it provides a chance to test equipment and identify areas that need improvement.

The ARRL SET is not the only opportunity that we have to test our emergency readiness. At any time throughout the year local SKYWARN groups can organize a drill or exercise. This is a particularly good idea to do just before a known storm season approaches. Here are a few suggestions for planning and conducting a drill or exercise for your group:

• Coordinate with local emergency management. Many local emergency managers have experience in setting up drills and exercises and can be a great resource. And during an actual activation storm spotters likely will work with local emergency management.

• Coordinate with the local NWS office and their SKYWARN coordinator. If they have an amateur radio station in the office see if it can be manned during your drill so you can practice sending traffic to the office via their station.

• Try to have a drill or test shortly after a SKYWARN training session. Interest among SKYWARN spotters will be piqued by the training and you will likely have more involvement in the test.

• Make drills and tests a regular event. Too much time between SKYWARN activities can end up in lack of interest.

• Try something new with each test. If you haven't used EchoLink or APRS or some other mode before, try it in your next test and see how it works.

• Follow up each test with an after-action report. Review what happened and make adjustments to plans as needed.

• Consult your local Emergency Coordinator, District Emergency Coordinator, Section Emergency Coordinator, and Section Manager for ideas. They may have experience in this area that can be helpful.

• When the drill is underway don't forget to announce clearly that it is a drill. Net control stations should make this announcement regularly.

• Should a real emergency situation arise during the drill, stop the drill using an announcement that all participants clearly understand such as "real world" or "real emergency." Identify a key stop phrase before the drill and make sure all participants know what it is.

• Don't forget to test emergency power sources such as generators and battery backups. This should be done for home stations, net control, and repeaters.

• And after the drill make a final report for your SEC and ARRL. This is usually done by the local EC.

Drills and exercises are not limited to ARES/RACES only. Your SKYWARN group can use these as methods to train members, improve response and communications, public relations opportunities, and ways to strengthen the tie between the NWS and the amateur radio storm spotting community.

Networking

Networking can be a valuable source of training and learning for the storm spotter. In many areas SKYWARN storm spotters are organized, through the local NWS office, into districts with district coordinators. Within each district there may be many local SKYWARN groups that activate during severe weather. Networking these groups together can be a tremendous asset. At the same time it can also be a tremendous challenge.

While it is not necessary to have a monthly area-wide SKYWARN meeting, it is a good idea to get area SKYWARN groups together perhaps on a quarterly or biannual basis. Topics for the meeting could cover a recap of activations, problems encountered during activation, assessment of communication between groups, talks on home weather stations, or arrange to have a NWS meteorologist there for a Q&A session. This meeting does not have to happen in person. Through technologies such as EchoLink, Skype, and wide area repeaters it is possible for SKYWARN groups to conduct a virtual meeting. If the local NWS office has these capabilities they can be networked into the meeting too.

Remember weather is dynamic; it can affect wide areas and involve dozens if not hundreds of storm spotters. We must also train to keep effective communication between SKYWARN groups going.

It is easy to think that training for storm spotters stops with SKYWARN training. SKYWARN training, basic and advanced, is the foundation for any storm spotter. These two courses are the "must haves" for the serious storm spotter. However, training does not stop there. There are countless resources to help us learn more about severe weather, safety, emergency communications, and preparedness. Our training only stops when we refuse to learn any more.

REFERENCES

[1]"Learn CPR – You Can Do It!": **depts.washington.edu/learncpr**

[2]For more information on the National Weather Service's role in promoting education refer to the NWS publication *National Oceanic and Atmospheric Administration Education Strategic Plan 2015-2035* available from **www.noaa.gov/education/explainers/noaa-education-strategic-plan**.

[3]JetStream — Online School for Weather: **www.weather.gov/jetstream/**

[4]National Hurricane Conference: **hurricanemeeting.com**

[5]*The Amateur Radio Public Service Handbook*, ARRL, 2012 (out of print)

Meteorology

The material in this chapter was prepared by Vic Morris, AH6WX

So what does a storm spotter look for? What is a reportable event? What hazards might a spotter face? The United States faces no shortage of severe weather events. Year round severe weather is a threat. Typically spotters are called on to observe severe thunderstorms, tornadoes, hail, flooding, damaging wind, and winter weather. But in some areas spotters may also be called on to assist with hurricanes and tropical storms, coastal erosion, dust storms, volcanic ash fall, extreme heat, fog, or surf conditions. In this chapter we will look at common severe weather events that may require storm spotter activation. We will look at tornadoes and funnel clouds, thunderstorms, hail, damaging wind, floods, and winter weather. These weather events are common throughout most of the United States and are reportable by most spotter networks. We will also look at other weather events that may require spotters to be active. We will look at basic information about each weather phenomenon, safety issues for the spotter, and reporting criteria.

Some areas of this country are prone to types of severe weather that others are not, such as hurricanes along the Gulf Coast. What each spotter group is activated for will depend on the types of severe weather their area. In fact SKYWARN training keeps this in mind, allowing each local NWS office to tailor training to meet local needs. But remember: just because hurricanes hit coasts, it doesn't mean other areas aren't affected. Hurricane Ike still packed hurricane force winds when it hit southwest Ohio, more than 700 miles inland! And while most deadly tornadoes impact the central United States, every state has experienced a tornado. In 2005 a tornado touched down near Sand Point, Alaska![1] While tornadoes in Alaska are rare, they do happen. Since 1950 only four tornadoes have been confirmed to have touched down in Alaska (three since 2004).

This chapter will provide the spotter with valuable information when observing severe weather. It will address some basic meteorological concepts behind the different weather events spotters may report on, reportable criteria, and some real-world accounts from spotters. Like the rest of this book, it is not designed to be a replacement for NWS SKYWARN training or other training provided by professional meteorologists; it is intended to complement the training that you receive.

THUNDERSTORMS

Thunderstorms are one of the most powerful and visible displays of natural energy. Many ancient cultures named gods after the fearsome power of lightning strikes. Even in our high tech age, lightning and other thunderstorm hazards are annually among the top natural disaster killers. SKYWARN storm spotters should have a good working knowledge of thunderstorm phenomena meteorology plus a high level of observational skills.

Before discussing the specifics of thunderstorms, let us investigate the processes of vertical motions in the atmosphere. This study will help one understand how many features of thunderstorms operate.

Suppose a parcel of air is forced to rise because it is crossing a mountain. The parcel will enter an environment of lower air pressure, and it cools due to expansion. If the volume of air is not fully saturated (no visible moisture present), it cools at 5.5° Fahrenheit per thousand feet, or 10° Celsius per kilometer. This is known as the *dry adiabatic lapse rate*.

If the parcel of air continues to rise, it will eventually reach *saturation*. This occurs when the parcel temperature and *dew point* become equal. At this stage the volume of air holds all the water vapor possible for a given temperature. Further cooling will reduce the parcel's ability to retain water vapor. The excess water vapor is condensed into liquid water, or perhaps ice crystals if the air is extremely cold.

When water vapor condenses it releases 600 calories of heat energy per cubic centimeter of liquid water formed. This effect is called the *release of the latent heat of condensation*. If the air parcel rises farther, the released latent heat will partially offset cooling by expansion. The net result is that the vertical cooling rate is slower. This is known as the *moist adiabatic lapse rate*. The moist rate is not linear depending on many atmospheric variables. Sometimes a simple approximation of 3°

Thunderstorm over New Mexico. [Craig Cornell, K9LHA, photo]

Fahrenheit per thousand feet or 6° Celsius per kilometer is used to help visualize the process.

Now suppose the air parcel passes the mountain crest and starts to sink. It will be compressed by rising air pressures. The parcel temperature begins to exceed its dew point and no visible moisture will be present. The air will warm up at the dry adiabatic lapse rate as it heads to lower elevations. Note this analysis assumes the air parcel is not interacting with the surrounding environment. Heavy precipitation occurs in many thunderstorms. Evaporation of the precipitation removes 600 calories of heat per cubic centimeter of water evaporated. This may cause an air parcel to become colder than otherwise expected. The importance of this will be discussed later.

Principles of buoyancy primarily determine the vertical extent of updrafts and downdrafts in convective clouds. If a rising air parcel is warmer than the surrounding atmosphere, it will rise like a hot air balloon on its own. If a parcel is not warmer than the surrounding atmosphere, it will be denser and tend to sink.

Meteorologists determine the existing vertical temperature, moisture, and wind distribution in the atmosphere with a *radiosonde*. This is an instrument package sent aloft by a weather balloon twice daily at hundreds of sites worldwide.

The observed vertical temperature pattern may or may not favor buoyancy of an air parcel. Suppose there is a layer of air in the atmosphere that is 80° Fahrenheit at its base. Now assume there is an air parcel of equal temperature at the base of this layer. Suppose 1000 feet higher the air temperature is 78°. Now let's force the air parcel to rise 1000 feet. It will cool by expansion to 74.5° Fahrenheit. Now remove the lifting force. What happens? The air parcel will sink because it is colder than its environment. This is an example of a stable air layer which is not favorable for buoyant updrafts.

There are rules for determining the stability of air layers. If the existing temperature decrease within the air layer is less than the moist adiabatic lapse rate, the layer is said to be stable. Note if the temperature actually rises with height, this is an inversion layer. Inversions are extremely stable. When they are close to the earth's surface, they trap pollutants near urban areas.

When the temperature difference from the base to the top of an air layer is between the moist and dry adiabatic lapse rates, it is said to be conditionally unstable. A rising volume of unsaturated air in this layer would not be buoyant. But if it is saturated it will arrive at the top of the air layer warmer than the environment. That means a saturated air parcel is buoyant in a conditionally unstable atmosphere.

If a layer of air cools vertically faster than the dry adiabatic lapse rate, the layer is unstable. Both saturated and unsaturated air parcels are buoyant in unstable environments. Unstable layers are usually shallow near strongly heated regions of the earth surface.

Now let's look at what happens in a typical vertically developed or convective cloud. One of a variety of reasons initiates upward motion of an air column. If the lifting process continues, the air reaches saturation forming a convective cloud base. This altitude is known as the *lifting condensation level (LCL)*. The height of this level may be estimated from surface temperature and dew point data.

The LCL height in feet is approximately: 222 (Temperature minus Dew point in degrees Fahrenheit).

For metric units the LCL height in kilometers is:

⅛ (Temperature minus Dew point in degrees Celsius).

But will the cloud grow? It may grow somewhat further if there is a sufficient mechanical force to overcome negative buoyancy. But to get a really well developed convective cloud, at some point the existing temperature pattern must allow rising air columns to become buoyant. The altitude where buoyancy is first achieved is known as the *level of free (or unassisted) convection or LFC*. Upon passing the LFC air columns will rise like hot air balloons. The air columns will keep rising until they encounter an environment warmer than the column.

In most cases, the lowest layer of the atmosphere — the troposphere — cools with increasing height up to 40,000 or 50,000 feet. Above this the stratosphere warms with increasing height, a form of inversion layer. Usually the top of the troposphere acts to restrain vertical cloud growth. The altitude where an air parcel or column runs out of buoyancy is called the equilibrium level. This is the theoretical convective cloud top, where a characteristic anvil-shaped cloud begins to spread downwind.

The actual altitude reached by buoyant air parcels depends on the range of altitudes where the parcel was buoyant plus the degree of buoyancy attained. These two factors strongly correlate to updraft speed. If there is a powerful updraft the convective cloud may overshoot the equilibrium level by thousands of feet. The amount a thunderstorm overshoots the equilibrium level is one clue in estimating its intensity.

Note that in the rising air column analysis just completed, no interaction with the environment was considered. Suppose there *is* some interaction. One example might be encountering a *rain shaft*. Cooling by evaporation could occur, diminishing updraft buoyancy. The reverse situation applies in downdrafts. The cooling will accelerate them.

Modern meteorologists have many means of assessing atmospheric stability and the potential for vertical motion. There are a wide variety of stability index products which are beyond the scope of this text. Interested spotters can learn more about them in many college level meteorology texts.

This discussion has strongly emphasized atmospheric stability. But what could change the existing or predicted stability? Any situation that increases the vertical temperature differences enhances instability. A very common source of unstable air is day time heating. Frequently the most unstable atmosphere occurs near or shortly following the daytime maximum temperature. A second instability source is cooling aloft. An approaching mid to upper level trough means colder temperatures at those levels.

A special note for anyone living in tropical ocean areas: there is very little day to night change in the surface temperature at sea. But the tops of existing clouds radiate more heat energy at night out to space. This means the tropical marine atmosphere is often most unstable late at night, resulting in the strongest convection during those hours. In contrast, surface cooling at night promotes atmospheric stability; an approaching mid to upper level ridge produces warming aloft, making the atmosphere stable.

Atmospheric stability conditions are a primary factor in determining thunderstorm potential. A second factor is the amount of moisture present. In extremely dry hot desert areas the atmosphere is often very unstable with turbulent strong updrafts and downdrafts. But there will be no clouds if the air is too dry.

Factors initiating updrafts (lifting mechanism) are important and have many varieties. Perhaps the easiest to understand is air forced to climb over rising terrain by the prevailing horizontal winds. Weather fronts divide air masses of different densities. A cold front vertically has a wedge shape which undercuts and lifts warm less dense air. Warm fronts have a gentle slope in which the less dense air glides over the colder air below. Ordinarily warm fronts are associated with layered and not convective clouds. The warm front discontinuity aloft is often an inversion layer. But if the warm air mass lifted is unstable and moist, gradual lifting may make air parcels buoyant. Embedded convection is possible with some warm fronts.

A third lifting mechanism, the dry line, often plays a key role in Texas and adjacent Southern Plains states. Suppose there is a general low level south to southwest wind flow during the warmer periods of the year. Air arriving from the Gulf of Mexico has a high water vapor content. Meanwhile air streams originating from northern Mexico will be hot but extremely dry. If the temperatures are similar, the dry air is denser and can act as a lifting mechanism and a focal point for new convection if conditions are otherwise favorable.

Small scale features are also important potential convection triggers. Near the Atlantic, Gulf of Mexico, and sometimes the Great Lakes, sea breezes may work their way inland on hot and humid days. These may trigger convection. Downdrafts from nearby thunderstorms act as mini cold fronts, and can set off new thunderstorm cells if the atmospheric conditions in the undisturbed air are otherwise favorable for convection.

Single-Cell Storm

Next we begin to describe and analyze the thunderstorm family. The simplest system is the *air mass* or *single-cell storm*. The sequence of events begins with a rising air column that forms a cumulus cloud.

If the atmosphere is relatively stable, the process terminates in a "fair weather" cumulus cloud that is only a few thousand feet deep. However, if the atmosphere is conditionally unstable through a deeper layer, the updraft may achieve buoyancy causing the cloud to quickly grow taller into towering cumulus — see **Figure 5.1A**. During this "cumulus stage" the cloud contains primarily updrafts.

As the cloud builds, vast numbers of water droplets collide with each other, forming larger droplets. Before long they become rain drops producing both showers and downdrafts. Meanwhile the portion of the cloud in rain-free areas keeps growing until nearing the equilibrium level, likely generating an anvil. Lightning and thunder are likely during this "mature stage," as shown in Figure 5.1B.

As the mature stage continues showers often become heavier and more widespread. Evaporational cooling near a vertical rain shaft main terminates buoyant updrafts nearby. The rain cooled ground makes the near surface layers more stable. Soon the updraft dies leaving only downdrafts in the "dissipating stage." The lower portions of the cumulonimbus cloud dissipate, leaving behind a remnant anvil for a while longer.

The processes described above for an air mass thunderstorm apply primarily when the winds aloft are not strong enough to horizontally displace storm updrafts and downdrafts. A significant percentage of thunderstorms fit into this category. Light mid to upper level winds are frequently found in the southern half of the United States during the peak of summer if no well defined weather front is near.

Multi-Cell Storms and Squall Lines

As previously discussed, thunderstorm downdrafts may act as mini cold fronts. These may initiate new thunderstorm cells near ones that have recently dissipated. There may be a number of cells next to each other simultaneously at various stages of development. Storms acquiring this pattern are called *cluster* or *multi-cell storms*. This is the most commonly observed type of thunderstorm.

The third type of thunderstorm organization is the *squall line* which is an extensive line of multi-cell storms. Squall lines are solid or broken lines of thunderstorms that may be hundreds of miles long. They often propagate ahead of a strong cold front, but squall lines may form in other weather scenarios. The line of storms may align nearly parallel to the approaching front. In other situations the storm organization is more complex. Squall lines can interact with terrain features, individual strong surface gust fronts, plus changes in the overall stability and upper level winds. If the shape of an approaching squall line acquires a bow shape, portions of the line may harbor damaging surface winds.

All types of thunderstorms may theoretically become severe. The National Weather Service defines a severe thunderstorm as one that meets at least one of the following criteria:

• Wind gusts of 50 knots (58 MPH) or higher;
• Hail 1 inch in diameter or greater;
• Any thunderstorm that produces a tornado.

Air mass and multi-cell storms forming in low shear (horizontal or vertical

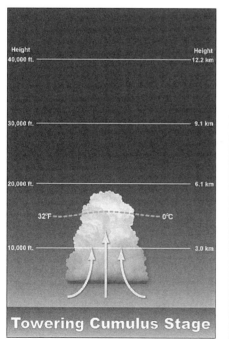
(A)

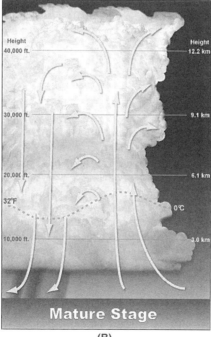
(B)

Figure 5.1 — (A) Developing stage of a thunderstorm. (B) Thunderstorm mature stage. [National Weather Service]

changes in wind speed and direction) rarely become severe. The NWS estimates that only 1% of all thunderstorms attain severe status.

Squall lines, however, tap into extra energy sources that increase the possibility of severe weather. They may generate small to moderately large hail, extensive downbursts (downdrafts reaching the ground), and some weaker tornadoes. Squall lines need to be monitored very closely by NWS personnel and SKYWARN spotters.

Factors favoring squall lines are warm, moist, unstable air masses lifted by a strong quickly advancing cold front. In the upper levels an approaching trough will lower mid to upper level temperatures enhancing instability. There is often considerable vertical wind shear as the wind speed increases rapidly with height. In rapidly advancing squall lines the updrafts and downdrafts usually tilt with height. Heavy rain and hail often fall outside buoyant updraft areas. That permits updrafts to continually build the storm. Meanwhile the leading edge of the downbursts / gust front will keep lifting surface air in advance of the squall line. This can regenerate a squall line for many hours and sometimes for a day or two provided the atmosphere remains favorably moist and unstable.

Supercells

The most dangerous type of thunderstorm is the *supercell*. Such storms are the most likely ones to produce violent tornadoes, huge hail, and the most intense downbursts. Updrafts in the most extreme supercells can exceed 150 MPH. Quite a few thunderstorms are buoyant enough to reach over 50,000 feet. Other storm varieties such as the squall line may have a long duration due to tilting updrafts and downdrafts.

But in a supercell rotation is an additional factor that enhances storm intensity. The rotation can be generated from two mechanisms. In the lowest few thousand feet of the atmosphere a strong, usually southerly, wind increases rapidly and veers with height. This feature is commonly called a *low level jet*. The wind shear from this wind pattern tends to generate a spin around a horizontal axis. Now suppose this spinning air is lifted by a strong updraft. The axis of rotation becomes more vertical and imparts rotation into the lower portions of an evolving supercell thunderstorm.

Meanwhile aloft a strong jet stream six to eight miles above the ground will cross the lower altitude rising columns of air at sharp angles inducing more rotation. When everything comes together properly, these two features will cause an entire supercell thunderstorm to rotate. The rotating updraft, accompanied by relatively low atmospheric pressure, is known as a *mesocyclone*. Mesocyclones pose a significant risk for tornado formation.

Once a supercell is established, it can be self-sustaining for many hours. A single supercell may generate multiple varieties of severe weather along its track. Note that the precipitation rate associated with a supercell is not an accurate indicator of the amount of severe weather the storm produces. As long as most factors remain favorable, a supercell sometimes persists overnight despite the usual night time surface cooling. The general public tends to think that tornadoes are mostly a daytime hazard. Late night tornadoes produced by supercells can be more deadly than normal if they catch people asleep and not prepared for them.

Thunderstorm Climatology

The annual frequency of all thunderstorm types varies according to season and location. A thunderstorm day is defined by the NWS as audible thunder heard.[2] In quiet rural environments thunder may be heard up to 25 miles away. Lightning can be seen up to 100 miles away at night from the tops of taller thunderstorms.

Annually most regions east of the Rockies experience 20 to 60 thunderstorm days per year. A majority of the storms occur during the spring, summer, and early fall, when the moisture and instability are greatest. A few Colorado and New Mexico eastern Rocky Mountain areas get up to 70 thunderstorm days per year. The north to south aligned mountain slopes are a significant source of lift when moist Gulf of Mexico air moves upslope.

From the Louisiana – Texas border to Georgia and points south, 70 or more annual days per year are also expected. The maximum frequency is just over 100 days per year in southwest Florida. The southern state locations with high thunderstorm frequency have a long season of moist and unstable air. Sea breezes near the coast often initiate air mass thunderstorms in summer.

Florida is a unique case with sea breezes arriving from the Gulf of Mexico and Atlantic in the northern and central parts of the state. Farther south sea breezes may arrive from the Florida Straits. The convergence of sea breezes is a very effective lifting mechanism that explains much of Florida's high thunderstorm totals. Very few of the thunderstorms started by sea breeze convergence become severe although sometimes they drop a great deal of rain. However, Florida can get severe thunderstorms associated with mid latitude strong cold fronts in the cooler months. Tropical cyclones in summer and fall may also generate thunderstorms meeting severe criteria.

Looking toward the West Coast, thunderstorm frequency is likely under-reported in sparsely populated mountain regions. The under reporting of the past is being corrected by modern lightning detection techniques. Continuing westward, a cold ocean current prevails along the West Coast. Normally the atmosphere is too stable for significant convection. But strong cold upper air lows can in some cases generate enough instability for coastal thunderstorms from late fall to early spring. A few summer thunderstorms can form over interior Southern California and Arizona when moisture flows northward from the Gulf of California.

Lightning

Outside of the mainland "lower 48" states, summer lightning strikes often ignite huge forest fires in Alaska. Hawaii gets thunderstorms infrequently, primarily associated with winter storms. In contrast the Commonwealth of Puerto Rico is a thunderstorm hotbed. I personally observed an average of 160 thunderstorm days per year from 1994 to 2001 in Rincon, a northwest coast town. Puerto Rico has an east to west mountain range that is very favorable for lifting moist and unstable air. Typically 50 to 75% of the rainy season (May to October) days will bring a thunderstorm somewhere to western Puerto Rico.

Over land areas, thunderstorms are most frequent from mid afternoon to early evening. This closely matches the times of expected maximum heating and instability. Over tropical oceans, thunderstorms are most likely late at night.

Lightning is a feature common to all types of thunderstorms. As a thunderstorm develops, varying parts of the cumulonimbus cloud acquire distinctly different electrical charges. Increasing updrafts and downdrafts bring opposite charged locations in proximity to each other. A lightning bolt discharges the electrical potential. Charge differentials between the cloud and the earth can lead to cloud-to-

Lightning is a common sight in the summer months. [Mike Corey, KI1U, photo]

ground lightning. The overall intensity of a thunderstorm is often quite well correlated with observed lightning flash frequency.

Lightning strikes can be deadly, and are a hazard to storm spotters as well as the general public. Studies have shown an average lightning bolt carries 30 kilo amperes of electrical current, and values exceeding 300 kilo amperes have been observed; currents as small as 100 milliamperes can be fatal.

Even relatively innocuous looking clouds can produce lightning. I will illustrate the point with a personal experience. When I was younger, I frequently went camping in North Carolina's Cape Hatteras National Seashore. One morning, around 6 AM, I awoke and noticed a towering cumulus cloud blowing in off the Gulf Stream. I wasn't concerned about the cloud, and I returned to my sleeping bag. Suddenly there was a blinding flash and a nearly instantaneous rifle shot crack of thunder. A solitary lightning bolt had struck a power line only 100 feet away. The cable sliced, leaving a dangerous sparking live wire on the ground. There was no other thunder or lightning produced by this lone storm cell.

In addition to general personal safety, lightning is a special concern to Amateur Radio operators. Towers and external antennas must be properly grounded. Hams living where thunderstorms are common should consider well designed lightning protection systems. And don't forget the wisdom of unplugging sensitive electronic devices before thunderstorms approach.

Thunder is the result of super heating of air by a lightning strike. The heated air expands suddenly and then contracts, creating a resounding crash of noise. The time interval between a lightning flash and a peal of thunder can be used to estimate the distance to the lightning. Simply measure the number of seconds between the two events and divide by five. This gives you the approximate distance to the lightning in miles.

But don't rely on this method to keep you safe from the threat of a lightning bolt. You should be practicing lightning safety rules whenever threatening and turbulent clouds are observed. It does not need to be raining at your location for lightning to strike. Bolts of lightning can travel miles through rain-free atmosphere. Strikes have been observed up to 10 miles away from the parent cumulonimbus cloud.

Numerous additional problems accompany severe thunderstorms plus some that do not technically reach "severe" criteria. Many thunderstorms of average intensity produce a short burst of heavy rainfall which is often beneficial. But on occasion a thunderstorm complex may move little for several hours. Or heavy precipitation cells may repeatedly cross the same region. This effect is known as "*training*." An unusually moist atmosphere is an obvious ingredient favoring excessive rainfall rates. The presence of an orographic barrier or stalled weather front can cause storms to repeatedly develop in the same general location.

Heavy Rain and Flooding

The result of a heavy rain varies greatly depending on rainfall rates, duration of the event, and the nature of the surface the rain strikes. Dry and sandy soil can absorb a lot of water before there is a potential flooding issue. Moist clay soils that are already saturated will absorb far less water before flooding becomes possible. Highly urbanized areas absorb almost no water, and runoff problems can develop very quickly in high intensity rains. The usual local problem in flat regions is flooding of streets, culverts, and low lying homes.

When the terrain gets steeper, runoff moves rapidly downhill, concentrating into torrents. This inundation, called a *flash flood*, can be extremely dangerous and may develop suddenly. In many years flash floods are the number one cause of weather related fatalities.

In the arid western states hikers sometimes camp out in relatively flat but dry old creek beds. If there are thunderstorms over distant mountains, sudden walls of water may sweep the campers to their deaths. The worst place to get caught is in a steep narrow canyon. Flash floods can rise feet in seconds and there is no means of escape.

Mid latitude storms can occur with or without frozen surface precipitation. Some are heavy rainfall producers leading to general river flooding.

To conduct a flood potential analysis, one must consider an entire drainage basin from the point of interest to all locations upstream. Previous studies available will determine soil types and their absorption capabilities. It is possible to estimate the degree of soil saturation. Detailed mapping will show the varying elevations throughout the drainage basin. Real time water levels are measured by stream gauges.

When a significant rain event begins, hydrologists need to evaluate numerous factors. Critical information includes how much rain has fallen over a drainage basin, the rainfall duration, and the rain intensity. The analysis must cover scales ranging from the smallest tributary to the largest river. Real time rainfalls are measured by a combination of rain gauges and Doppler radar estimates. These combined with a dense network of surface observations, augmented by SKYWARN spotter reports, provide "ground truth" that can overcome some Doppler radar limitations.

When a flood event occurs forecasters will evaluate a combination of real time data and numerical forecasting models. Appropriate warnings sent to local authorities permit timely evacuations and flood mitigation measures.

Flood forecasting in colder locations is especially challenging during winter thaws or the normal spring warming. The depth and water content of snow must be evaluated for an entire drainage basin. Very accurate temperature forecasts are needed to determine the snow melting rate. This rate of melting partly determines how quickly the released water reaches streams and rivers. The water moves most quickly over solidly frozen ground or other impervious surfaces. If the ground has thawed it may be able to absorb some water if the soil profile is not saturated.

Unexpected situations can alter flood forecasts. There is always a risk of a serious precipitation prediction error. Ice jams can act as dams creating unexpectedly high water upstream until the ice breaks up. Or man-made levees could fail. When a levee is breached, the flood covers a wider area, but the overall average water level is lower.

Hail

Hail accompanies quite a few stronger thunderstorms. Annually it is responsible for roughly $1 billion worth of property and crop damage. Even relatively small hail can damage or destroy crops. Larger hail stones dent motor vehicles, shatter glass, and cause personal injuries.

On occasion hail can get surprisingly large. Every year there are reports of baseball to softball size hail. The largest measured hail, *seven inches* in diameter, fell in Nebraska during 2003. This chunk of ice weighed almost a pound. A slightly smaller but denser hail stone from Kansas in 1970 tipped the scales at 1.7 pounds.

Hail typically falls in swaths or streaks that are considerably narrower than the source thunderstorm cell. Frequently hail or hail mixed with rain persists only a few minutes. Once in a while persistent hail can accumulate several inches deep, creating a winter-like landscape out of season. Heavy rain accompanying hail or nearby flash floods can push large hail accumulations around, creating locally much deeper hail measurements. High winds in some hail events can also create hail drifts.

Hail initially forms when an updraft lifts rain into below-freezing portions of a cumulonimbus cloud. The chunk of ice can grow due to collisions with super-cooled droplets. These are tiny water droplets in a cloud that may remain unfrozen at temperatures below 0° Celsius. Super-cooled droplets freeze on contact with larger snow or ice masses in the storm cloud. In the case of hail these droplets produce *rime*, or cloudy-appearing ice.

Developing hail sometimes escapes the source downdraft and gets coated with liquid water where the temperature is above freezing. Then the hail may get recaptured by an updraft. The rain freezes on to the hail stone adding a layer of clear ice. Hail stones may cross the freezing altitude several times, growing as they transit updraft and downdraft regions. Mature hail stones frequently display alternating layers of clear and rime ice. The shape of large hail varies widely often displaying irregular protruding lumps or spikes.

At some point hail stones become too heavy to be supported by the existing updrafts, and they fall to the ground. Note that very large baseball size hail stones may fall at terminal velocities near 100 MPH. The fall speed can be faster if assisted by strong downdrafts: Large hail may also fall through strong horizontal winds. This causes the hail to strike exposed objects at an angle.

Hail at any one location is not a common event. In North America hail is most frequent in the Great Plains region, but it is possible in any state. The higher elevation Plains states are especially hail-prone because the freezing level is often only a mile to a mile and one half above the ground. The hail stones have minimal opportunity to melt as they fall. Stations in warm humid climates near sea level have freezing levels at least three miles above the surface most of the time. The tropical coasts get comparatively infrequent hail (less than once per year) since most of it melts before reaching the ground. Many tropical regions rarely experience the thunderstorm types most likely to cause large hail.

Damaging Straight-Line Winds

Thunderstorm downdrafts frequently intersect the earth's surface and then spread outward. In milder thunderstorms the cool gust front provides welcome relief from a hot and humid day. In more intense thunderstorms the gusts may easily exceed the severe thunderstorm limit of 58 MPH. Every year there are documented reports of straight line winds well over 100 MPH.

The rush of air from a thunderstorm is known as a *downburst*. If the swath of damaging wind is 2.5 miles or wider, it is considered to be a *macroburst*, while swaths narrower than 2.5 miles are considered to be *microbursts* (**Figure 5.2**).

In most cases initial downbursts persist for only a few minutes. The most violent wind speeds occur just after the downburst intersects the surface. Wet downbursts associated with heavy rain often display a strongly sloping rain shaft close to the ground. Dry downbursts, accompanied by minimal or no rain, raise areas of blowing dust over open country.

The violent winds in downbursts can down trees, endanger mariners, and cause considerable property damage. Sometimes in squall line thunderstorms and adjacent macrobursts merge causing lengthy damage swaths as the squall line advances. These events are sometimes called *derechos* (**Figure 5.3**). *Derecho* is a Spanish word for right. The name was given because many of the prolonged

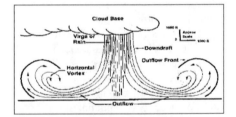

Figure 5.2 — Diagram of a microburst. [National Weather Service]

Straight-line wind damage. [Bo Prince, photo]

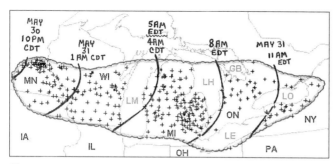

Figure 5.3 — Progression of a *derecho*, May 1998. [National Weather Service]

downburst events tend to deviate to the right of the course taken by their parent storms. Prolonged macroburst events are most frequent in the Great Plains region, but they are not too uncommon during the most thunderstorm prone months from the Midwest, to the Northeast, and Mid Atlantic regions.

A lot of people fly regularly, and they should be aware that microbursts are a special concern for pilots about to land. Normally on final approach a pilot flies a gradual descent on a predetermined glide slope angle. During descent the pilot gradually loses forward speed, planning to touch down near the end of a runway. The prime factor determining the lifting force of an airplane wing is the air speed relative to the aircraft motion. If a pilot encounters an approaching microburst head on, the wind pushing his airplane up (air speed) increases suddenly. The pilot finds himself well above the intended glide slope. An improperly trained pilot may take steps to lower his plane to get back on the desired glide slope. This action can be a fatal mistake. Suppose the pilot emerges on the other side of the microburst. A lot of things can go bad very

A Pilot's Take on Wind Shear

There are two types of wind shear: *increasing performance* and *decreasing performance*.

Increasing Performance Wind Shear

Imagine a plane is sitting on the ground. The plane is not tied down and the wheels aren't chocked. Now a big wind kicks up. If a plane were light enough and the wind hit the wings straight on, the wind could cause the wings to create enough lift to make the airplane lift off the ground without the engine running.

Now put the plane in the air. The plane is flying along though stable air. Then, the plane flies into an area of weather and the wind is shearing in all different directions, like a frontal zone or a gust front. If the direction of the wind changes and causes a quick increase in wind on the nose of the plane the plane will also experience an increase in the airspeed of the plane. The result is an increase in lift and/or airspeed, but most likely both will increase at the same time.

Deceasing Performance Wind Shear

The anatomy of a downburst or microburst is air moving straight down and once it hits the ground it spreads out in every direction. Imagine water being poured into a basin. As you fly into the leading edge of the microburst you will always get increasing performance. In the middle the burst will cause an airplane to lose altitude because the air mass the plane is flying through is moving vertically downward (again imagine the water pouring out of a bucket). As the plane flies out the other side of the downburst the headwind has quickly become a tailwind. For example, a 50 knot head wind shearing to a 50 knot tailwind results in a loss of 100 knots of wind on the nose and therefore an equivalent loss of lift. This decreases the performance of the wing and can cause a plane to stall. This happened to Delta Flight 191 at Dallas-Fort Worth in 1985 and the plane was lost during approach.

In the event of a downburst, depending on where you are relative to the down/micro burst, you can experience one or the other or both. I experienced this flying into Amarillo (AMA) a few months ago. There was a cell on the far west side of the field and I was approaching from the southeast. With no warning my airspeed began increasing at an alarming rate. There were some visual clues, such as a cloud formation which would suggest a gust front. In addition to the increasing airspeed, the turbulence was moderate in intensity. My airplane has a wind speed and direction indicator. It quickly changed from a headwind of about 25 knots to a 35 – 40 knot crosswind from the left (so it changed from the northwest to the southwest) in just about one mile. It was probably my most exciting landing.

The other experience was on take-off from White Plains, New York (HPN) on a winter's day. There was a very strong wind from the south. We were taking off to the east. The airport is surrounded by trees which work as a shield to the wind waiting above the trees. Once the airplane was airborne and flew above the trees, the turbulence began at moderate intensity. Then, the EGPWS (Enhanced Ground Proximity Warning System) which also handles wind shear alerts gave me a visual and aural warning. The next thing I heard was "WINDSHEAR, WINDSHEAR." The captain was the PNF (pilot not flying) and told me it was increasing performance, and to ride it out (we are trained to shout out our vertical speed trend and distance above the ground until we are clear of the wind shear. Because it was increasing performance we did not have to worry about the ground too much, we were being shot away from the ground at high speed. My initial level off altitude in my clearance was 2000 feet MSL. That is about 1600 feet AGL. Because of the turbulence and difficulty to maintain control, I was unable to control my altitude and exceeded my clearance limit by 1000 feet before I could reverse the climbing trend. In the event at Amarillo, I never got a wind shear alert, which surprised me given my airspeed increased about 40 knots in a few seconds.

I have never experienced decreasing performance wind shear, thankfully. I have trained for it, and have practiced wind shear penetration in the simulator. In the training exercise the airplane has good airspeed, but loses altitude because the air mass is moving vertically downward and not horizontal. In these cases, we are trained to select APR Power (Automatic Power Reserve), that is an aircraft specific term which spools the engine to maximum power, then we pitch the aircraft to the "Pitch Limiter" and try to fly out of the wind shear straight ahead. Once out of the wind shear, the PLI (pitch limit indicator) will disappear and then the pilot has to bring airspeed and climb rate into control. There is no doubt that an Air Traffic Control clearance limit will be broken, but the Federal Aviation Regulations state that the pilot in command may deviate from the FARs to meet any emergency. Also, as I stated before, that while the PF (pilot flying) flies the plane through the wind shear, the PNF will shout out vertical speed trends, and height above ground from the radar altimeter. — *Craig Cornell, K9LHA* quickly. The head wind will suddenly shift to a tail wind, and there will be much less lift on the wings. The plane will drop well below the safe glide slope, possibly leading to a crash unless corrective maneuvers are taken at once. These involve increasing power and attempting to escape the microburst with course changes. To make matters worse, the pilot may have just seconds to react during violent turbulence and forces pushing downward on the airplane.

A microburst documented near Washington, DC almost involved Ronald Reagan when he was president. *Air Force One* had just landed. Six minutes later a microburst beyond the active runway caused the wind to speed up from 15 MPH to over 140 MPH in two minutes from one direction. Then briefly the wind speed dropped almost like in the eye of a hurricane. Next the winds jumped well over 100 MPH from the opposite direction. Finally they simmered down to normal speeds a few minutes later.

It is doubtful any airplane can survive such an extreme microburst if it catches the aircraft at the wrong moment. Fortunately such an encounter is much less likely now. Pilots are well trained on microburst flight techniques. Doppler radars can detect many potential downbursts, especially larger ones. Many thunderstorm prone busy airports have also installed low altitude wind shear alert systems (LLWAS) as a last defense to keep airplanes out of microbursts.

TORNADOES

Tornadoes are the "top gun" in the severe thunderstorm arsenal. Despite their awesome power, research meteorologists still have a lot to learn about these fearsome phenomena.

The most dangerous tornadoes usually form in connection with supercell thunderstorms. The starting point is usually a mesocyclone, an area of low pressure and strong updrafts, best developed at mid levels. Falling pressures at the surface tend to cause the mesocyclone to expand downward. The descending mesocyclone is likely to contract, causing it to spin more rapidly. This potentially can lead to a funnel cloud and then a tornado. Meanwhile an adjacent downdraft, often known as a *rear flanking downdraft*, also descends to the ground.

Once the tornado touches down, it can intensify as long as there is an inflow of warm, moist air to support the parent mesocyclone. Eventually cool downdrafts wrap around the tornado, cutting off the supply of buoyant air. When this occurs, the tornado will weaken and dissipate.

Note that an intense supercell mesocyclone may rejuvenate multiple times if it finds new sources of moist and unstable air. As the parent thunderstorm advances it can be responsible for a series of tornadoes. The threat of additional tornado developments sometimes persists for hours until the supercell finally weakens.

Not all tornadoes form in connection with supercell storms. Sometimes converging low to mid level streams of air associated with other thunderstorm types acquire sufficient rotation to spin up a tornado. In squall lines tornadoes have a tendency to form on the southernmost cell or in cells just north of a break in the line. Tornadic rotations can also briefly occur in multicell storms.

The visual appearance of tornadoes and

Tornado seven miles south of Anadarko, Oklahoma, May 1999. [NOAA library]

Tornado damage in Tuscaloosa, Alabama, May 2011. [Mike Corey, KI1U, photo]

related clouds are well documented in NWS SKYWARN spotter training materials. You may wonder what causes a funnel to appear in the first place. As you realize, a tornado is a microscale area of intense low pressure. Air that moves into lower air pressure cools by expansion. This expansion process is identical to the one in rising air parcels. If the cooling is sufficient, the air will cool to the dew point, and condensation will occur. In relatively dry air masses, there may be no easily visible funnel until the circulation contacts the surface, raising dust and debris.

Tornadoes vary greatly in size, intensity, and path length. An average tornado is less than a quarter mile in diameter, lasts 10 minutes, and is on the ground for a few miles. A tornado tends to travel in tandem with the parent thunderstorm. In the United States a high percentage of twisters approach from the southwest or west. A typical "Tornado Alley" type storm has a forward speed near 35 MPH, but the speed range can be from stationary to over 100 MPH. Tornado tracks are not always straight. They may swerve suddenly in response to changing air currents nearby. In rare cases tornadoes have even reversed course, such as the one near Dimmitt, Texas in June 2005.

Tornado intensity is now measured by the *Enhanced Fujita Scale*. The range is from EF0, a gale force tornado, to EF5 which means three second gust speeds above 200 MPH. The tornado intensity is determined by NWS post storm survey. The estimated strength is based upon the damage produced accounting for the strength of construction materials.

The most violent three categories of tornadoes account for less than 10% of those observed in the United States. The top category EF5 accounts for less than 1% of storms. A graphic example of EF5 effects was the near total destruction in Greensburg, Kansas in May 2007.

Nobody knows exactly the maximum possible tornado wind speeds or the minimum air pressures. However, a DOW (Doppler radar On Wheels) once measured a 302 MPH peak gust in a tornado near 100 feet above the surface in Oklahoma. An instrument package deployed in advance of another approaching tornado survived to record a 100 millibar pressure change in seconds as the twister passed.

The path length of a few large tornadoes may exceed 50 miles. The all time credible record was established by the 1925 Tri-State tornado. It was on the ground for 219 miles, as it spent four hours crossing portions of Missouri, Illinois, and Indiana. This tornado produced more documented fatalities (at least 695) than any other to date in our history.

Tornado shape, diameter, and general appearance can vary rapidly. At maturity some larger tornadoes are over a half mile wide. The record width based on storm damage surveys appears to be near 2.5 miles. Very wide tornadoes are sometimes difficult for the general public to recognize in the distance. They may appear to be huge turbulent masses of black cloud whose rotation may be obscured by vegetation, terrain, or man-made features. Large twisters are sometimes called *wedge tornadoes*.

In the majority of cases, there is a steady progression from a funnel cloud to a tornado ground contact. As the tornado strengthens it sometimes gets wider. During the weakening phases a dissipating funnel gets narrower and sometimes bends.

There does not appear to be a strong correlation between tornado diameter and peak wind speeds. A relatively narrow tornado can be extremely strong. Larger tornadoes often have several small vortices rotating inside the main funnel. Sometimes these are easily observed by storm spotters. The small vortices may account for some of the strange damage patterns seen in a tornado passage. Suppose there are two equally well built homes across the street from each other. After the storm one house may have relatively minor damage because it encountered 100 MPH winds. Across the street a mini vortex brought an additional 100 MPH making the total wind speed 200 MPH. The unfortunate second home is probably totally destroyed.

Tornado climatology is important information. All 50 states have tornadoes, but the relative frequency is highest in the "Tornado Alley" states from Texas to the North Central Plains — see **Figure 5.4**. Extreme air mass contrasts are frequently present, setting in motion the factors that generate twisters.

A second high frequency area is Florida. The peninsula is a unique case. It is sometimes exposed to supercell thunderstorms in the cooler months of the year. Passing tropical cyclones may spin up tornadoes in their spiral squall bands. The three way convergence of sea breezes may generate multicell thunderstorms capable of brief weak tornadoes.

In general most winter tornado formation is confined to the southeast Atlantic region plus states bordering the Gulf of Mexico. Only these spots are likely to see occasional episodes of the required warm, moist, and unstable air. As spring approaches strengthening solar radiation more strongly heats the ground in the southern states. Meanwhile snow and very cold air retreats slowly in higher latitudes. Tornado frequency picks up over the Southern Plains and interior southeastern states. By April and May most of the United States east of the Rockies can see the recipe for tornadoes. Activity peaks in the Great Plains states, but violent tornadoes can form farther east. The April 1974 Xenia, Ohio tornado is a good example.

By summer the jet stream retreats northward, and there is often insufficient wind shear in Tornado Alley to produce the most violent storms. The northernmost states from the Dakotas to New England often have their greatest tornado frequencies during June, July, and sometimes August.

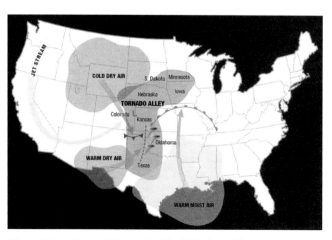

Figure 5.4— Map of "Tornado Alley." [National Weather Service]

The Joplin Tornado

Missouri is no stranger to tornadoes, and the state lies along the eastern edge of the historical high risk tornado belt. 2011 was an extreme year for US tornado outbreaks. The all-time most twisters in a single outbreak came in late April across the mid-southern states. May was not to be denied as ingredients came together for another major tornado outbreak of 241 twisters spanning May 21 – 26. Several notable severe tornadoes struck during this time, but the Joplin, Missouri tornado of May 22 was catastrophic.

Meteorological conditions pointing to a Plains severe weather outbreak on May 22 were noted by the Storm Prediction Center as early as May 20. SPC assigned a moderate risk of severe storms to a region that included Missouri. Warm, moist, unstable tropical air was expected to move northward from the Gulf of Mexico ahead of a slow moving cold front. Aloft a trough was predicted to send a jet stream on a track over the southern Plains. This would provide the higher level vertical wind shear conducive for severe thunderstorms. However, computer forecast models did not show much wind shift with increasing height in the lower atmosphere. Thus the early guidance emphasized large hail as the most serious threat although tornado activity was considered to be possible.

The outlook changed little until the morning of May 22 when the SPC noted the highest probability of thunderstorm development was from Oklahoma to Missouri. Low level wind shear was still considered to be marginal but more than sufficient for tornadoes given extreme instability. The Springfield, Missouri Weather Forecast Office took a special radiosonde observation at 1900 GMT (1 PM CDT) to evaluate conditions. At 1:30 PM CDT SPC issued Tornado Watch 325 covering Missouri and adjacent states. "Explosive thunderstorm development" was likely and "a strong tornado or two is possible."

The first thunderstorms of the day developed in Kansas between 2 – 3 PM and rapidly became severe. These prompted a number of severe thunderstorm warnings while moving mostly east. The first tornado warning of the afternoon was issued at 4:25 PM. Additional thunderstorms formed near the Kansas-Missouri border near 5 PM, and these also quickly became dangerous, identified as having a high tornado risk. A tornado warning was issued on one cell at 5:09 PM, and the warned area included the north part of Joplin in the southern part of the Tornado Warning #30 polygon. Sirens were sounded in Joplin at 5:11 PM, but no tornado from this warning struck Joplin.

A second Tornado Warning, #31, was issued on a storm west of Joplin at 5:17 PM. Storm spotters reported a tornado touchdown at 5:34 PM. At first the tornado was small diameter and produced EF0 to EF1 minor damage. However as the twister entered southwest Joplin it intensified dramatically to EF4 and became a wide wedge-shaped tornado half a mile to a mile in diameter. This development took only four minutes. The tornado sirens sounded for three minutes a second time at 5:38 PM. Spotters now reported sighting a damaging, multiple vortex tornado.

The southern third of Joplin was mostly destroyed as the tornado tracked generally east at 20 MPH with slight deviations. The path of EF4 to EF5 (over 200 MPH gusts) damage was six miles long and a half to three quarters of a mile wide through the city of population 50,000. Homes, businesses, medical buildings, and other structures all had overwhelming damage. Many homes were blown entirely off their foundations. Steel trusses were seen rolled up like paper. Some 300 pound concrete parking lot stops anchored with rebar were torn out and tossed 60 yards. The St. John's Medical Center lost windows, many interior walls, ceilings, part of a roof, and a life flight helicopter. Later the building was torn down as it was deemed structurally unsafe. Semi rigs were tossed 125 yards and wrapped around debarked trees. After the tornado left Joplin it weakened, but it didn't dissipate entirely until it completed a 22 mile track

The death toll was 158, the first tornado with more than 100 fatalities since the Flint, Michigan, tornado of 1953. The Joplin tornado was the seventh most deadly tornado in United States history. There were at least 1000 injuries. Damage was estimated at $2.8 billion, the costliest tornado on record. Emergency responders had a Herculean task with the large area of death and catastrophic destruction.

The National Weather Service Central Region conducted an intensive investigation in the months following the Joplin tornado. There were many significant conclusions, but only a few are discussed below. First forecasters overall had an excellent grasp of the weather events leading up to the tornado outbreak. However there was never language designating a PDS (particularly dangerous situation) tornado watch as computer guidance was not calling for a massive tornado outbreak on May 22. The Doppler radar in Springfield, Missouri, couldn't detect the lowest portions of the Joplin tornado, and there was only one volume scan per 5 minutes. As a result forecasters weren't aware of the sudden intensification of the tornado as it entered Joplin. There was no tornado warning language telling people of the exceptional danger in this storm.

Other factors contributed to the high number of fatalities. There was some confusion caused by multiple tornado warnings. Some people regarded the warnings with complacency, taking an

As fall approaches coastal states may get tornadoes produced by tropical storms and hurricanes making landfall. Otherwise tornado frequency tends to drop as the solar heating diminishes. However, in some years there may be a secondary tornado maximum in October and November. This is possible if surface storm tracks and jet stream positions are similar to those seen in spring.

Waterspouts observed over the ocean and sometimes lakes are a subspecies of the tornado family. There are two basic types. The "fair weather" waterspout usually forms where surface waters are much warmer than the overlying atmosphere. This creates very unstable conditions near the ocean or lake surface. Now introduce a towering cumulus cloud and some wind shear near the surface. The shear may be produced by converging sea breezes or outflows from nearby decaying convective storms. As the air rises funnels may form in this environment. The typical fair weather waterspout is associated with towering cumulus or weak cumulonimbus clouds. Most of the time the water spouts are EF0 to EF1 intensity.

In certain circumstances supercell and squall line thunderstorms may move over lake or ocean waters. If these thunderstorms generate a waterspout it may be stronger than many fair weather waterspouts. The waterspouts originating from intense cumulonimbus clouds are called "tornadic waterspouts." Waterspouts are not counted when the NWS compiles their annual tornado statistics. However, if a waterspout of either type moves onshore, it is classified as a tornado and counted.

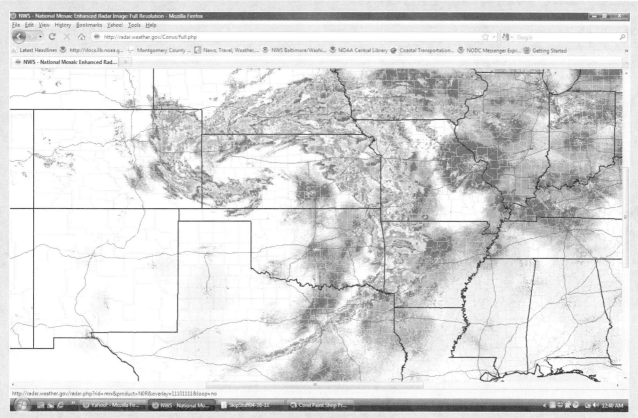

NWS enhanced radar image of the weather system that spawned the Joplin Tornado.

"it can't happen here" attitude. More than 50% of the fatalities occurred in residences. Few homes in Joplin have basements or tornado cellars. The advice to move to an interior room on the lowest floor does little good if a tornado is strong enough to blow away an entire structure. The NWS provided a lead time of 17 minutes warning on the Joplin tornado. This was slightly better than a recent average warning lead time for violent tornadoes. However many of the fatalities were elderly, infirm, or disabled persons. There may not have been sufficient time to move them to a safe place.

Meteorologists have been very concerned what might happen if a violent tornado moves through a major city. It is only a question of when not if this will happen. The Joplin tornado of May 22, 2011, provided insight about what can happen. Better preparation on all levels is the key to tornado survival. — *Vic Morris, AH6WX*

References

NWS Central Region Service Assessment: Joplin, Missouri Tornado-May 22, 2011, National Weather Service, Central Region Headquarters, 2011. No authors listed.

Storm Chaser Home Page: www.twisterchasers.com

U.S. Tornadoes: www.ustornadoes.com

Wikipedia: en.wikipedia.org/wiki/2011_Joplin_tornado

Waterspout off the Florida Keys, September 1969. [NOAA library]

VALUABLE WEATHER TOOLS

SKYWARN storm spotters are called on to report on more weather conditions than thunderstorms and their related events. Shortly we will look at some of these other weather phenomena and some things that fall more into the realm of geology and other earth sciences. Before we do so, we should discuss two of the most valuable tools to anyone interested in weather, weather radar and weather satellites.

Weather Radar

Modern Doppler weather radar provides many ways to evaluate weather systems of interest to SKYWARN spotters. Briefly I would like to review the history of weather radar. Radar as a military application was invented prior to World War II. Personnel on Oahu, Hawaii actually observed incoming aircraft from the Japanese Pearl Harbor raid in 1941 without realizing what they were seeing on their radar screen. After the war ended, the official position of the United States Weather Bureau (USWB) was to not mention the word tornado in forecasts because of the risk of causing panic (the USWB was the precursor to the National Weather Service). At the time, severe storm forecasting was in its infancy, and we did not have any means to detect tornadoes except by direct sightings.

The USWB recognized the value of radar in detecting precipitation, and the WSR-57 radar was deployed at many USWB offices during the late 1950s and 1960s. The early radars could only measure precipitation intensity, altitude, and range.

I had personal experience forecasting in the pre-radar period. Fast forward to a January night in 1971. I was a junior weather officer at the Navy Fleet Weather Central in Pearl Harbor. Ironically 30 years after the Pearl Harbor raid, there was still no weather radar in Hawaii. The only computer forecasting guidance available to Navy forecasters was the appropriately-named "primitive equation model." Forecasters did the best they could with surface observations, intermittent polar orbiting weather satellite coverage, plus some radiosonde reports. The closest one was from Lihue on Kauai which is west-northwest of the island of Oahu.

That particular January night I was on a 12-hour mid watch. In Hawaii most of the time we experience east-northeast trade winds with dew points in the 60s. However, at midnight there was a stiff south-southwest wind and a dew point over 70°. I knew from the surface analysis chart a strong low pressure area was passing a few hundred miles north of the Aloha State. The local air pressure was falling rapidly.

As part of my duties, I was responsible for preparing a local Hawaii forecast for US Navy and Marine facilities. As I was considering how to word my forecast, the latest Lihue sounding came in. Much to my surprise the report indicated extremely unstable air plus a vertical wind profile strongly supporting severe weather. Just one problem: at that time neither the Navy, Air Force, nor USWB forecasters had procedures for generating severe thunderstorms warnings.

So I went ahead and engineered a local forecast that called for torrential rains, frequent lightning and damaging shifting winds without actually saying the word "severe." Perhaps I exceeded my authority as a lowly Navy ensign. Luckily I was right or I could have been in serious trouble with my commanding officer. There were reports of 60 – 90 MPH wind gusts later on Oahu. After sunrise, a tornado injured a few people in a shopping center at Kailua-Kona on the Big Island of Hawaii.

The NWS embarked on a modernization program as computer technology developed and the old WSR-57 radars became obsolete. Now the WSR-88D Doppler weather radar systems and upgrades cover all 50 states, Puerto Rico, and Guam. Deployment of the Doppler radar has greatly improved short term detection of many varieties of adverse weather conditions.

The Doppler radar emits very short pulses of UHF energy that produces reflections from precipitation sized drops, snow, ice, and so on. The radar operator can adjust the sensitivity of the unit to meet his needs. The clear air mode is ultra sensitive and can detect unexpected items such as insects. Most events of interest to storm spotters are observed when the radar is in the precipitation mode.

The second objective for Doppler radar is to determine the motions of precipitation and local winds with respect to the radar. Doppler motion is detected by noting the frequency changes in reflected signals. To understand the Doppler principle, imagine an approaching whistling train. The sound appears to increase in pitch as both the sound waves and train are headed toward the observer. After the train passes, the sound is travelling in the opposite direction from the train. The observer hears a lower pitched whistle.

Modern Doppler radar systems can very accurately determine the velocities of particles directed toward and away from the site. Numerous applications of Doppler wind data are used for severe thunderstorm, hail, and tornado analysis. If a tropical cyclone is close to the radar site, the wind distributions can be accurately mapped out. There are additional Doppler applications for winter weather.

When the WSR-88D is operated, the radar dish may be elevated through 14 different angles from 0.5° upward to a maximum of 19.5°. The elevation choices permit the radar to probe clouds at various altitudes and distances from the radar site.

There are six NWS Doppler radar displays readily available to all users on the Internet. Other displays are available for those with special software, and you may see them from time to time on *The Weather Channel* and similar programs. I will cover the basic Doppler displays and their capabilities in this section. Later I will cover limitations that Doppler radar users should recognize.

Base reflectivity is perhaps the most commonly selected presentation. It shows the returns detected when the radar transmitter is elevated at a 0.5° angle. Composite reflectivity displays the highest intensity return on any of the 14 scans between 0.5 and 19.5° in the vertical. Short range displays of each reflectivity can reach up to 143 miles in favorable circumstances, and the long range reflectivity scan has a theoretical detection range of 286 miles.

The reflected returns are a color coded decibels (dBz) display. Blue and green are typically at the low end of the intensity scale. Moderate precipitation rates are typically yellow to orange. Heavy to extreme rates are typically red, purple, or white. Larger hail stones often display as extreme precipitation rates. Hail echoes often display very sharp, slightly irregular shapes. This contrasts to more gradual and variable shape changes in the display around heavy rain areas. Large hail will trip alarms in the local NWS Weather Forecast Offices.

A second set of displays shows the relative winds observed. A Doppler radar only can measure winds toward or away from the site along the 0.5° elevation of the radar beam. Green colors are typically a "toward" site wind indication, and red are typically an "away" measurement. To estimate the actual wind speed and direction intercepted by the radar beam one must know exactly where the radar site is located. The Doppler radar only shows full wind speed if the wind is blowing at right angles to the Doppler. As the wind direction deviates from the perpendicular, the presentation will shows winds lighter than reality. Some trigonometry and vector analysis can still provide good wind speed and direction estimates in the area of interest. If the wind direction is parallel to the Doppler, the operator will observe no velocity indication.

NWS Doppler radars can operate in the storm relative mode; this display compensates for storm motion. To illustrate, suppose you are on the bow of a ship moving 20 knots (nautical miles per hour) into calm air. You will feel 20 knots of wind. If there is a wind of 10 knots moving directly at you, the relative wind is now 30 knots. This is similar in concept to the microburst produces aircraft air speed changes described in an earlier section.

The storm relative Doppler display is extremely useful for locating the inflow and outflow regions near convective storms. Although the display usually lacks sufficient resolution to see tornadoes, it frequently picks out precursor Mesocyclones.

A "red flag" display is a couplet of small scale adjacent/away velocities embedded in an otherwise consistent velocity field. When the toward/away difference exceeds a preset value, alarms are triggered indicating a *tornado vortex signature* (*TVS*). If other indications suggest tornado potential, the local WFO will issue a tornado warning. Not every TVS results in a confirmed tornado, but forecasters are steadily improving their success rates and lead times. Recent policies now permit forecasters to issue warnings for only the portions of counties they feel to be at high risk. In the past most severe warnings had been issued for entire counties, creating an unacceptable false alarm rate for many residents, even if a small percentage of them did experience severe weather events.

The final two public Doppler radar displays are hourly rainfall and a forecaster selected storm total rainfall. While these displays are useful, they should be taken with a few grains of salt, especially if the estimate concerns long distance precipitation. Unfortunately Doppler radar has some significant limitations on its accuracy.

The first issue is generated by the altitudes of the radar beams (0.5 to 19.5°) and the curvature of the earth. If you live less than 20 miles from the radar, the radar beams may not reach some local precipitation (the cone of silence), and the precipitation rate may be underestimated. You may get a more accurate picture from a somewhat more distant radar. In addition near the radar man-made structures or natural terrain may reflect some of the radar energy cone. This false return is called *ground clutter*. Computer algorithms can cancel out some ground clutter effects.

Typical flat terrain Doppler radars can detect the majority of precipitation within 80 miles and intense precipitation within 140 miles from your location. Sometimes at medium ranges radar will detect cloud level precipitation that evaporates before reaching the surface. As the distance to precipitation bearing clouds increases, the radar will intercept only the tallest portions of the clouds. The operator might see low intensity returns from cloud tops that are actually producing plenty of precipitations at altitudes under the radar beams. Satellite interpretation skill is a high priority where radar coverage is limited.

When a radar beam intersects very heavy precipitation, the energy beam is weakened before it can intercept more distant rain, snow, or hail. The more distant bad weather will show weaker than normal returns on the radar screen. This effect is called *precipitation attenuation*.

If you live near mountainous terrain, you are at the mercy of your local Doppler radar altitude. A significant range of horizontal directions or azimuths may be blocked off by higher ground. Ground clutter could obscure showers falling on the closest mountain slopes. Showers forming below the 0.5° upward radar beam may be undetected at considerably lower elevations. And you may not be able to see what is happening beyond the mountain crest either. But recall that the composite precipitation mode shows the strongest returns from any of the 14 radar dish elevation angles. That means the composite radar has a good chance of detecting heavier precipitation areas missed by the base level scan.

Anomalous propagation or super refraction occurs when strong inversion layers are present. The radar beam may be bent back to earth in unexpected locations. This causes false returns where no precipitation is occurring. On stable windy days some coastal radars get persistent returns from sea spray.

Perhaps worse, the bent radar beam may miss areas of real precipitation. If in doubt the best advice is to seek a second opinion. Refer to the latest surface observations and satellite imagery. As good SKYWARN spotters you are already carefully observing your local sky.

One last radar issue is worth mentioning for those exposed to winter weather issues. Heavier snow aloft will gradually melt into rain as it reaches elevations above freezing. The partially melted snow often gives a false indication of heavier precipitation known as a bright band. Checking local surface observations is the best means to determine the precipitation rates in these cases.

Weather Satellites

Weather satellite imagery has greatly enhanced our ability to analyze worldwide weather and make far more accurate predictions. SKYWARN spotters can gain a great deal of weather insight by learning some basics of satellite meteorology. The history of satellite meteorology has provided some interesting and exciting moments.

German scientists developed long range rocket capability near the end of World War II. During the subsequent Cold War, a space race developed between the Soviet Union and the United States. Space became a new frontier for military and civilian applications. The Russian satellite *Sputnik*, launched in 1957, was followed by the United States *Explorer I* satellite a year later.

Meteorologists recognized the potential of having an eye in the sky above the atmosphere. In 1959 the first weather satellite was launched, *Vanguard II*. It was designed to provide information on cloud cover and density in the atmosphere. Technical problems resulted in poor quality from the optical instrument on board. The *TIROS I* weather satellite launched in 1960 exceeded all expectations and became the basis for future weather satellites. Weather scientists quickly found out that many weather features are well organized and easily tracked from space.

As the 1960s progressed, civilian polar orbiting weather satellites were launched. These provided relatively low resolution visible and infrared images. But the low frequency of satellite passes meant that weather features could go undetected for 12 hours or longer.

Meanwhile the military engaged in

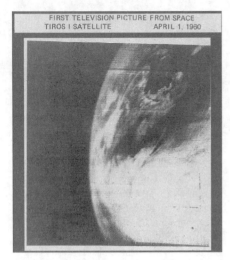

Image from TIROS I. [National Weather Service]

some ultra top secret programs to keep track of our Cold War enemies. Very high quality and high resolution images were generated by a secret satellite. This satellite played a key role in an incident involving *Apollo 11*. After our astronauts returned from their triumph on the moon, they were scheduled to splash down in a capsule in the tropical Pacific southeast of Hawaii. President Richard Nixon planned to be onboard the Hornet to greet our returning space heroes. Captain Hank Brandli, USAF, at Hickam Air Force Base had access to the very tightly compartmentalized ultra top secret imagery. He could clearly see that a tropical disturbance with 50,000 foot tall cumulonimbus clouds was heading into the *Apollo* recovery zone. If the astronauts parachuted into such weather, they risked death, and the United States risked a national disgrace. The young Air Force captain desperately needed to contact somebody to share his forecast information.

The Fleet Weather Central at Pearl Harbor was responsible for the Apollo recovery area forecasts. Fortunately the commanding officer Captain Willard S. Houston, USN, had a security clearance that allowed him to see Brandli's data. Brandli appealed to Houston to do something. Captain Houston approached Rear Admiral Donald Dixon of the *Apollo* recovery group. The admiral did not have the clearance to see the secret images. And it would be a security violation for Captain Houston to disclose sensitive top secret information. But Captain Houston very persuasively passed on the warning to the admiral, despite a risk to his career.

Rear Admiral Dixon took up the matter with his Washington superiors. Fortunately the top military brass and senior NASA officials listened. The *Apollo* re-entry trajectory was altered, and the recovery task force was moved to a new safe location. A potential tragedy was avoided, and later Captain Houston was awarded a medal for doing the right thing.

The story above was declassified by President Clinton during the 1990s. The polar satellite data involved, the Defense Military Satellite Program (DMSP), is now readily available on the Internet.

Let's talk about modern weather satellite systems. As already noted, one common trajectory is a polar orbiter. The satellites typically take nearly 90 minutes to complete one orbit. The earth rotates under the satellites, causing the sensors to look at different locations on each orbit. That somewhat limits their ability to track weather unless a number of polar orbiters are available. On the plus side, polar orbiting satellites can follow trajectories just above the outer atmosphere, enabling them to obtain very detailed data. They are also a benefit to high latitude locations such as Alaska.

Geostationary satellites are fired into low latitude eastward orbits that match the rotation rate of planet earth above the equator. The satellite altitude is near 23,000 miles. At that altitude the satellite appears to hover above a fixed point on earth. The geostationary GOES and equivalent satellites can continuously monitor and track the weather below. The large distance from earth permits a GOES "bird" to sense almost an entire hemisphere. But the data resolution is only moderately detailed. Given their consistent location above the equator, these satellites have less value to high latitude locations because of the earth's curvature.

Many varieties of data can be transmitted by both polar orbiting and geostationary satellites. Basically the satellites observe outgoing radiation at different wavelengths of interest to forecasters.

Visible imagery is the easiest to understand. Imagine you are on a spacecraft observing the earth below. You see the tops of cloud patterns, but not what is underneath them. Strong convective clouds are normally easy to identify as they are often taller than other clouds present. The prime limitation is that one is unable to observe the weather where it is dark.

Infrared imagery senses the temperature of the cloud, water vapor, or ground surface radiating energy toward the weather satellite. The data

Visual and infrared images of Hurricane Irene (2011) from NOAA-18. [Courtesy Joe Carcia, NJ1Q]

may be processed in numerous ways, resulting in temperature coded color pattern displays. Some color schemes brightly display the coldest and presumably tallest cloud tops.

Others enhance only cloud temperatures falling in specific ranges. Shortwave infrared often depicts shallow to medium depth cloud patterns not well displayed by some conventional infrared displays. Water vapor radiates well on specific wavelengths, and modern imagery is good at detecting regions of moist and dry air at various levels.

The strong point for infrared imagery is 24-hour coverage. Rapidly cooling cloud tops are a sign of strong vertical motion and thunderstorm development. Land and sea surface temperatures can be measured if no clouds are present. Such data is an important input for weather forecasting models.

Infrared does have one shortcoming. Suppose a cloud temperature is close to a surface temperature. In such cases it may be difficult to tell what the satellite is sensing. An example of this issue is detection of fog and other means need to be used to tell if fog is present.

There are some special weather satellite applications that deserve some attention. A modern generation of satellites contains microwave sensors. These are able to penetrate cloud canopies and detect high intensity rain cores. Microwave data can reveal hurricane eye wall structures, enabling more accurate hurricane intensity estimates. Other microwave sensors can estimate the surface winds at sea and approximate wave height.

OTHER WEATHER EVENTS

Storm spotters may be called on to observe and report a wide range of weather events and non-weather events that can be related to weather. Some of these events, such as dense fog, may occur just about anywhere. Other events such as volcanic ash fall are isolated to specific locales. Warning Coordination Meteorologists (WCMs) at NWS Weather Forecast Offices have considerable latitude in the types of information they may request from spotters. This allows the WFO to customize products to fit local needs. Spotters near the coast could be asked to report coastal flooding, beach erosion, and surf heights. In this section we will look at a few of these events and how they are important to the storm spotter.

Winter Weather

In this section we will discuss many of the impacts of non convective winter type storms. To better understand how such storms do their dirty work, it is important to briefly review air mass theories and the life cycles of middle latitude low pressure areas. These are often given technical terms such as the *wave cyclone* or *extratropical cyclone*. A cyclone is any type of low pressure area. An extratropical cyclone derives its energy from differing air mass densities. In contrast, *tropical cyclones* exist in a uniform air mass. They draw their energy from the release of vast amounts of latent and direct heat energy obtained from warm ocean waters.

The fundamental cause of *all* weather is unequal heating of our planet by the sun. Large expanses of air covering millions of square miles acquire similar temperature and humidity. These are air masses. Descriptive terms such as *continental polar* or *maritime tropical* give a snapshot of expected conditions.

When two contrasting air masses are next to each other, things get interesting. Cold dry air is dense, and it will attempt to undercut less dense warm humid air. In physics terms, there is potential energy awaiting release.

Suppose there is a stationary boundary of significant air mass contrast oriented east to west. And assume that the winds are blowing from 180° different directions on opposing sides of this discontinuity or weather front. What will release the potential energy?

The answer is gained from a vertical view in the troposphere. Suppose there is a region of divergence (air moving away from a location) in the upper atmosphere. The weight of the air column below lessens causing falling surface pressure. Air will rise to compensate for the air diverging at the top of the column. This is a recipe for bad weather. The falling pressure results in counterclockwise rotation of air around a point on the stationary front. Note if there is convergence aloft, the air sinks and surface pressure rises.

The wind speed normally increases with height in the middle to higher latitudes, although there are exceptions. The strongest winds are concentrated in a tube of air six to eight miles above sea level, called the *jet stream*. The jet stream is located above regions having a high horizontal temperature contrast. The polar jet stream is typically located over the middle latitudes, moving seasonally with the path of the sun. A secondary subtropical jet stream sometimes develops between 15° to 35° of latitude.

Meteorologists are especially interested in jet stream cores of maximum wind speed called jet *streaks*. Note that jet streams speeds can top 200 MPH. When a streak is viewed from above, the shape is often somewhat elliptical. The longest axis is aligned parallel to the wind maximum. Looking along the path of a jet stream the ellipse may be divided into quadrants in relation to the spot with peak wind speeds. Ahead and downwind of the jet maximum, the quadrants are labeled left front and right front. Behind the jet maximum the quadrants are labeled left rear and right rear.

Observations and theory demonstrate there is a consistent pattern of divergence and convergence associated with each quadrant. Divergence and upward motions are associated with the left front and right rear jet quadrants. If these pass over a stationary front, a well defined low pressure center may form. The right front and left rear quadrants are convergent. If these pass over the same stationary front, a significant low pressure area is not likely, but other forms of bad weather may still persist; however, if a new low pressure area develops, there will be increasing counterclockwise winds around it in the Northern Hemisphere. Frictional effects cause the wind to cross the isobars (lines connecting points of equal air pressure) slightly in the direction of lowest air pressure. As the wind picks up, the potential energy of the air mass contrast is converted into the kinetic energy of wind motion.

The surface fronts begin to move with respect to the low pressure center which tracks along with mid level steering currents. Advancing cold air undercuts warm air, creating a *cold front*. Warm air glides over retreating cold air as a *warm front*.

The zone between the surface cold and warm fronts is called the warm sector.

Usually the cold front advances faster than the warm front, and the warm sector gets narrower. If conditions at all levels are favorable, the low will get stronger and wind speeds will increase. Eventually the cold front catches up with the warm front. This leads to an *occluded front*. In an occlusion surface cold air on opposing sides of the front meet. Remaining warm air is forced aloft. The air mass horizontal density difference greatly weakens. The low is usually near peak intensity when occlusion begins. As the occlusion process continues at increasing distances from the low, the entire circulation gradually spins down.

The mid latitude storm tracks and intensities follow seasonal jet stream cycles. During the colder months the jet stream is usually stronger and farther south. Passing storms can affect almost any state depending on day to day weather pattern shifts. In the warmest months the jet stream is weaker and often not too far from the border between Canada and the United States. Mid latitude storms are most likely to affect the northern half of the country.

Precipitation

The first section of this chapter concerned primarily weather hazards generated by *convective clouds*. This section emphasizes problems produced by *layered clouds*. To really understand precipitation, one must look at microscale processes within clouds.

In the previous section it was assumed that condensation occurs when air is chilled to the dew point. But that is only part of the story. Moisture needs something to condense *on*, called a *condensation nucleus*. This is a very tiny particle of natural dust, sea salt, or man-made pollutant. Some nuclei such as salt actually *attract* moisture and are called *hygroscopic*.

The average cloud droplet is 20 microns in diameter, although there are significant size variations (a micron is one millionth of a meter). Then what causes precipitation?

The simplest case to analyze is a cloud entirely warmer than freezing (32° Fahrenheit or 0° Celsius). Both horizontal and vertical winds within the cloud cause the varying sized droplets to collide. Sometimes the droplets bounce off each other. In other cases they merge or coalesce. Cloud physics research shows that droplets containing opposing electrical charges are most likely to merge. As the droplets get bigger and heavier, they fall faster and merge more frequently with their neighbors. Showers will soon reach the surface. In some cases rain can fall only 15 to 20 minutes after a warm cumulus cloud forms.

All year in the tropics, plus sometimes during the middle latitude summer, freezing levels are near three miles above sea level. Rain that falls from clouds entirely warmer than freezing is referred to as "warm rain." Studies have shown that cumulus clouds as shallow as a few thousand feet deep can produce showers, especially if they contain salty hygroscopic nuclei.

Many clouds contain regions that are both warmer and colder than freezing, and numerous winter clouds are entirely below freezing. Such clouds may contain different varieties of water droplets and ice crystals depending on the vertical temperature distribution. In cloud regions warmer than 0° Celsius, the expected conventional water droplets form. But in clouds below 0° Celsius there is an unusual form of liquid water called *super cooled droplets*. Most people think of the freezing point of water as 0° Celsius. That is entirely correct for large volumes of water such as a lake surface or even ice cubes in a glass. But at the diameter of tiny cloud droplets, the water can remain as a liquid down to temperatures well below 0° Celsius. If one inventories a cloud at temperatures below 0° Celsius, there will be a mix of super cooled droplets plus some ice crystals. The fraction of ice crystals increases as one gets closer to -40° Celsius. Most of the naturally occurring freezing nuclei "activate" at about –10° C at which point the cloud consists entirely of ice crystals.

When falling ice crystals contact super cooled droplets, the water freezes immediately. The ice crystals grow at the expense of moisture in the previous super cooled droplets. The growth of a falling ice crystal is known as *accretion*. Once sufficient growth occurs, the complex of ice crystals becomes a snowflake.

As snowflakes enter regions above freezing beneath the cloud they will melt into a cold rain. The process of rain formation that began with ice crystals is known as the *Bergeron process*.

Snowfall

The majority of snowfall occurs in connection with passing mid latitude storms. Quite often during winter the warm sectors of such storms will be warmer than freezing. But as the low pressure area moves along its path, warm air will be lifted gradually ahead of the surface warm front. Snow is likely from east through northwest of the developing low center if the entire region is colder than freezing. If only some portions of a winter type storm are colder than freezing, the situation can get complicated. I will illustrate the points with some examples from New England, where I grew up. The examples provided here can be applied elsewhere in similar cases.

The first hypothetical storm directly approaches Cape Cod, a peninsula off southeastern Massachusetts, from the southwest. The open ocean near Cape Cod is usually 35 to 40° Fahrenheit during the winter. If the wind is from the east, any early snow on Cape Cod will quickly turn to rain. Meanwhile away from the coast in interior Southern New England there will be a swath of heavy snow. Farther northwest a light powdery snow tapers off well away from the storm center.

In scenario #2 the low center moves across central Massachusetts. Locations east of the low will see primarily rain. West and north of the storm track expect mostly snow, although other forms of frozen precipitation are possible. I will cover those later. In many cases the heaviest snow occurs 50 to 100 miles northwest of the low pressure center track. Once the low center moves by, the winds turn northwest, and it gets much colder. Areas of slushy wet snow will freeze creating lots of slippery ice. Snow squalls are possible in areas that received mostly rain. I used to call mixed precipitation followed by a freeze a "slop" storm. There are likely unprintable terms for the same mess!

In scenario #3 the low center passes 50 to 100 miles east of Cape Cod. If the air near the Cape is cold enough, it will be their turn to get a heavy snow. Locations farther inland will receive smaller totals.

Certain locations near water bodies such as the Great Lakes are susceptible to a special type of snow storm. The lakes become relatively mild in summer, and are slow to give up their heat as the weather gets colder. Suppose a very cold arctic air mass crosses an unfrozen large lake. The air is heated from below, and a layer of very unstable air develops. Lines of cumulus clouds build, and these may become very prolific snow makers. Quite often persistent low level wind directions cause lines of heavy snow bearing clouds to train over the same location for many hours. There is often a sharp divide between minimal to excessive lake effect snowfalls. Prolonged lake effect events can pile up to

Winter storms and nor'easters can cause severe damage to antennas. [Dave Patton, NN1N, photo]

several feet — just ask any resident of Buffalo or Cleveland.

Snow storms are possible in portions of all 50 states. As expected the northern United States cities usually get the largest annual total snow falls for heavily populated areas. But snow flurries have been seen as far south as Miami. In Hawaii the taller peaks of Mauna Loa and Mauna Kea on the Big Island plus Haleakala on Maui get snow during most winters.

Freezing rain and ice pellets (*sleet*) deserve special attention. Suppose there is a slow moving warm front or one that has turned stationary. At higher altitudes the air glides up the frontal slope. Suppose some of the rising air is warmer than freezing. At your surface location it is below freezing. Rain develops initially in the warm air aloft. But if it falls through a considerable depth of below freezing air, it will freeze creating small relatively harmless chunks of ice called ice pellets.

If the near surface below freezing layer is shallow, the rain will not have a chance to freeze in the air. Instead it freezes on contact with all surfaces colder than 32° Fahrenheit. The result is very dangerous freezing rain. It quickly covers streets and encases power lines and tree limbs. Prolonged freezing rain events can easily produce over an inch of ice. That is a recipe for lengthy power failures, downed branches, and extremely hazardous driving conditions.

Given the proper temperature patterns, certain weather or geographical situations favor persistent freezing rain. A nearly stationary front may be aligned across a portion of the eastern or central United States. If the jet stream position is not favorable, no major storm center will form. Instead a series of weak lows may move along the front causing moderate lift. Surface locations on the colder side of the stalled front may get many hours of freezing rain.

Mountain ranges such as the Appalachians often act to dam cold air masses ahead of approaching warm fronts. The warm air may not be able to dislodge cold dense air in mountain valleys. These are locations very susceptible to freezing rain.

Let's look at winter weather hazards on a broader scale. The obvious problems of a heavy snowfall include difficult driving, closed schools and businesses, and prolonged flight delays. Matters get far worse if there is high wind.

The NWS defines a blizzard as snow or blowing snow accompanied by frequent 35 MPH or stronger gusts which lower visibilities to ¼ mile or less. Motorists caught in a blizzard can be stranded for many hours. Drivers in blizzard-prone areas should have winter weather survival gear in their vehicles. In blizzards snow drifts many feet tall form, and high winds can redeposit snow almost as quickly as plows can remove it. Hikers and others outdoors in open country can get disoriented and lost. Prolonged exposure may lead to frostbite and possibly death.

The high winds associated with a blizzard are produced by strong pressure gradients near the parent low pressure center. If a strong Arctic high pressure area moves in behind the departing low, blowing and drifting snows are likely well after the snowfall ends. During this post storm period rapidly falling temperatures are common. The mix of high wind and cold air creates dangerously low wind chill readings. Most residents of cold climates know they should dress in layered winter clothing, and limit their exposure to very low wind chill temperatures.

Nor'easters

Intense coastal mid latitude storms create additional hazards. Along the United States Atlantic coast, a fairly common cool season storm track is from near Florida to the Carolinas and New England. These are often called "*northeasters*" by Atlantic coast residents. Many of the storms reach gale to storm force (39 – 73 MPH), and a few may exceed hurricane force (74 MPH or greater). Very intense storms also frequently approach Alaska, Washington, Oregon, and Northern California during the same months.

Usually — but not always — the highest winds are at sea. If the storm moves rapidly, the main issue near the coast is a few hours of wind strong enough to cause minor damage. However, very high winds at sea will create waves large enough to endanger commercial fishing boats.

On infrequent occasions a very intense mid latitude storm can stall or move slowly on erratic path for days. The classic example of this is the "Perfect Storm" of 1991. In the closing days of October that year, an unusually strong Arctic high pressure area set up over northern Quebec. It sent blasts of very cold air over the still relatively warm Gulf Stream south and east of New England and Nova Scotia. An extremely potent upper level trough created severe divergence aloft along a front southeast of Sable Island, Nova Scotia. Surface pressures fell very rapidly around an explosively deepening storm. Additional energy was also provided by the remnants of a late season hurricane moving into the area.

By October 29 a well defined 970 millibar storm was established south of Nova Scotia. The air pressure was 75 millibars higher in Quebec. The great pressure

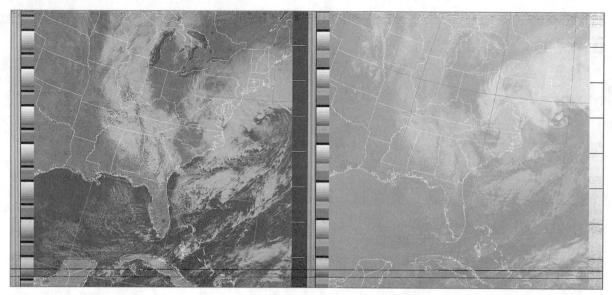

Nor'easter Athena occurred shortly after Hurricane Sandy. Visual and infrared images from NOAA-19. [Courtesy Joe Carcia, NJ1Q]

difference created a very large region of storm to hurricane force winds over the northwest Atlantic. Usually middle latitude storm systems move toward the northeast or east. But the extremely strong upper trough near the storm caused the system to make a slow counterclockwise loop over several days.

The maximum wave heights observed at sea are explained by three factors. These include wind speed, duration of high wind, and the *fetch length* (a fetch is a region having relatively constant wind speed and direction).

During the "Perfect Storm" there were unusually strong winds blowing for days over hundreds of miles of ocean. Extraordinarily high seas developed. A Canadian buoy south of Nova Scotia recorded a peak wave 101 feet high. This is the largest one ever measured by a weather buoy. The exceptionally high seas caused the fishing boat *Andrea Gail* to sink. Several other boats were also lost or severely damaged.

The winds were near hurricane force in gusts along the eastern New England coast. Storm tides (the sum of the astronomical tide plus storm surge) were several feet above normal. They may have reached 10 feet above normal on parts of outer Cape Cod. At Truro on the Cape, the huge seas broke through a 100-yard wide, 20 foot high dune to create a new ocean inlet. The inlet remained open for several months. Numerous New England shoreline properties were battered by waves and flooded.

Very large swells created huge surf in areas far removed from the storm. Almost unprecedented 20 foot breakers reached Palm Beach, Florida. Some 35 foot waves were seen off northwestern Puerto Rico.

The Pacific coast is exposed to large waves more frequently than the Atlantic coast. Winter storm tracks quite commonly aim potentially dangerous fetches at Hawaii, California, Oregon, Washington, and Alaska. Battering waves and the run up of ocean water periodically threaten coastal properties. Very large waves produce deadly rip currents that can pull swimmers out to sea. Never walk along a beach backed by steep wet cliffs during high surf. That is a warning the waves have reached a high elevation, and you could be caught by a breaker with no place to escape.

Beach erosion is a related issue impacting coastal communities. A prolonged period of heavy surf will lower the beach profile depositing the removed sand well offshore. As the beach gets lower, waves can attack the dunes behind them, perhaps undermining some structures. The worst beach erosion often occurs when a persistent gale is blowing across the coast at an acute angle for several consecutive high tide cycles. During one storm in the 1980s I personally observed a 100 foot width of beach disappear at New Smyrna Beach on the north central Florida Atlantic coast. This phenomenon will likely only be exacerbated in the coming decades due to the observed sea level rise attributed to global warming. After the storm some lucky residents found treasure coins and other artifacts from old Spanish shipwrecks.

Dense Fog

Fog is a weather problem that affects many people, and it can be highly localized. On occasion freeway drivers traveling at 70 MPH suddenly enter pockets of near zero visibility. If they fail to slow to a safe speed, multi vehicle pileups are very possible. Spotter reports can help forecasters better pinpoint locations for dense fog advisories.

There are several major fog types. Suppose at sunset the sky is clear but it is humid with little wind. Heat energy radiating from the surface will eventually cool the air to its dew point temperature. The result is *radiation fog*. It tends to pool in low spots, and it is often most widespread late at night through the early morning hours.

A second variety of fog is common near coasts with cold water. During the warmer months humid air blowing over the chilly oceans will get cooled to its dew point. This process creates *advection fog*. Advection means the horizontal motion of some atmospheric property. Both the New England coast and the Pacific coast in North America have high advection fog frequencies. Advection fog tends to thin out or dissipate by day over land. But it will likely persist just offshore. After sunset the coast cools off and the fog will likely return.

If humid air climbs a mountain range, the expanding air may cool to its dew

Early morning radiation fog. [National Weather Service]

point. *Upslope fog* is the result. Note an observer at a lower elevation will be looking up toward a cloud base.

During winter very cold and dry air can cross unfrozen waters. Moisture evaporates from the lake or ocean raising the cold air dew point; when the dew point and temperature become equal, fog forms. The technical name for the result is *evaporation fog*. But frequently this fog variety is referred to as "*Arctic sea smoke*," a very accurate description of its appearance.

NON-WEATHER EVENTS

Landslides

Steep slopes are prone to problems related to both weather and geology. Suppose there is a prolonged period of heavy rain falling on a deep layer of poorly consolidated soil. If the soil gets saturated, portions of a hillside may give away suddenly, causing a mudslide or a landslide. Areas devoid of natural vegetation due to recent wild fires or poor grading practices are particularly prone to sudden collapse. In steep rocky areas heavy rain can cause rock slides. These also can cause property damage and are a safety hazard.

Avalanches

Very deep accumulations of snow are common in mountain regions. If the snow overhangs a steep slope, it can collapse suddenly, starting an avalanche. Avalanches become very likely when snow gains weight from the seepage of nearby melting. There is almost no advance warning for a specific avalanche. Skiers and snowmobile drivers may be threatened. Ski patrols sometimes deliberately set off avalanches in controlled safe environments to reduce the risk.

Wildfires

Another significant hazard is fire weather. Every year numerous very large wildfires occur in our country. A long previous period of dry weather precedes most major fires. This situation is a normal part of some climate types. For example, Southern California normally gets very little rain from mid spring to mid fall.

The fire hazard greatly increases on dry windy days. Red flag warnings are issued by local WFOs when wind gusts are high and the relative humidity is low. The fire threat will be especially high if there are any lightning strikes into dry vegetation..

Many large scale weather scenarios can create conditions favorable for wildfires. One of the most notorious is the Southern California Santa Ana wind. The basic setup is a strong high pressure area over the Great Basin near the fall season; air blowing clockwise from the high pressure area heads toward the California coast. As the air sinks it warms by compression and the relative humidity sometimes drops to single digit values. To make things worse the winds can accelerate near mountains and canyons. Sometimes the gusts top hurricane force. Out of control blazes can spread very rapidly leaving very little time to evacuate. The risk to human lives has increased in recent years as more people build homes in high risk wildfire zones.

Volcanic Activity

Weather Forecast Offices in the Pacific coast states sometimes need to deal directly or indirectly with geological phenomena. These locations lie along the Pacific "Ring of Fire," and volcanic eruptions are possible. Volcanoes tend to fall into two varieties. Most from California to Alaska tend to be moderately to highly explosive due to gas trapped under pressure in subsurface molten rock known as magma. The Mount St Helens eruption of 1980 serves as an outstanding example. Such eruptions not only produce lava, but they may also generate pyroclastic flows — superheated dense masses of volcanic ash and poisonous gases which can move down the slopes at speeds well over 100 MPH.

Volcanoes covered by heavy snow and glacial ice present additional hazards. If lava or a pyroclastic flow reaches deep ice, sudden melting will unleash tremendous flooding. Things get even worse if a crater lake breaches adding to the torrents. The sudden deluges contain volcanic ash, mud, and debris. These events are known as *lahars*.

Mount Rainier east of Seattle is a prime candidate for future lahars. Several hundred years ago geological evidence indicates that lahars dozens of feet deep reached some current cities of southwestern Washington State. Repeat performances are likely in the future from Mt Rainier, and perhaps other glaciated Pacific coast volcanoes.

Volcanic ash can spread considerable distances based on vertical wind profiles. Falling ash is a serious health hazard that impairs breathing function. Heavy ash falls can lead to roof failures and other property damage. The problem of roof collapse is exacerbated if the volcanic ash becomes wet from rain as it takes on the consistency of concrete. Suspended volcanic ash is a serious danger to aviation. The ash can be ingested by jet turbine engines. Engines can abruptly cut out and shut down, resulting in power loss and rapid loss of altitude. If a pilot cannot find reasonably clear air to restart the engines, a crash is likely. Accurate aviation forecasts can enable pilots to avoid ash areas. Ash issues were caused in 2009 by Mt Redoubt in Alaska as well as eruptions in the Northern Mariana Islands, a United States western Pacific possession.

Hawaiian volcanoes are located over a Pacific "hot spot," and they behave differently from their "Ring of Fire" cousins. Violent eruptions are extremely rare, and usually lava flows are comparatively gentle. In the past, glowing hot lava has devoured several villages on the Big Island of Hawaii. Kilauea volcano, in eruption since 1983, sent streams of lava into the Pacific Ocean near the Hawaii Volcanoes National Park as recent as 2009. Adjacent much taller Mauna Loa volcano erupts less frequently, but it has the potential to send lava toward some heavily-populated areas.

Although not currently destroying homes, Kilauea has been a bad neighbor for surrounding communities. Kilauea's Puu O'o and Halemaumau craters have released 1500 – 3000 tons of sulfur dioxide gas per day since mid 2008. Depending on local wind conditions this natural pollution source can reach all parts of the Big Island and sometimes the rest of the state. One by-product of the sulfur dioxide release is particulate pollution. The poor resulting air quality can burn sensitive crops, aggravate breathing problems, and greatly reduce visibility.

Tsunamis

Tsunamis are a hazard most commonly associated with the Pacific Ocean. Historical research shows that approximately 59% of destructive tsunamis have occurred in Pacific waters. While that is the case, other oceans are at risk. This fact was demonstrated by the catastrophic Indian Ocean tsunami just after Christmas in 2004.

There is a long list of very destructive tsunami events recorded in history. While the Pacific leads the list, people living in the Atlantic and especially the Caribbean region should not be too complacent. During 1867 a magnitude 7.5 earthquake occurred in the Anegada trough near what are now the United States Virgin Islands.

People running from the tsunami in Hilo, Hawaii, April 1, 1946. [United States Geological Survey]

Local tsunami waves following the quake were estimated to reach up to 40 feet high. These tossed ships on land and destroyed port facilities. That is those still left. Three weeks earlier a Category 3 (111 – 130 MPH) hurricane had damaged many existing structures.

In 1918 a similar earthquake took place off northwest Puerto Rico. The seismic waves reached at least 20 feet above sea level. I lived there 80 years later, and I could still observe remnants of tsunami destroyed lighthouses, and exceptionally sharply angled sea cliffs cut by the tsunami.

In November 1929 an estimated magnitude 7.2 earthquake shook the Grand Banks area near Newfoundland. Rushing tsunami waves entered fiord like inlets and rose more than 50 feet in a few locations. Fortunately the population was sparse in the worst affected areas. Minor tsunami effects were recorded as far away as South Carolina.

It took a long time to learn how to predict an ocean-wide tsunami. A pivotal event happened on April 1, 1946. Around 1 – 2 AM very strong earthquake displaced the sea floor just beyond Unimak Island in the Aleutians. Not long afterwards a wall of water at least 100 feet high washed away the Scotch Cap lighthouse. The ocean reached an estimated 115 feet above sea level. Seismographs on Oahu recorded the event, but nobody knew that a Pacific-wide tsunami had been generated. Tsunami waves smashed into the Hawaiian Islands just as people were getting ready for work or school. The wave run up heights varied greatly around the state, but most north shores had heavy damage. The tsunami run up ranged from 30 to 55 feet above sea level in the most exposed locations. The waterfront sections of Hilo on the Big Island were demolished. 159 people lost their lives.

I lived briefly on the north shore of Oahu at Sunset Beach during 1971 – 1972. I got to know a 90-year old gentleman who often walked down a nearby path to watch the waves and surfers. One day he told me his 1946 tsunami tale. He was casting nets to catch fish when suddenly the water began to drain off the reefs. Trouble coming! He ran as fast as his then 65-year old legs could carry him toward the Koolau slopes that start ¼ mile inland of the beach. But he could see he wasn't going to beat the incoming wave. So he climbed a tall coconut palm, and waited out several successive tsunami surges. When the event ended, there was a sea water lake left in the lowest spots near his palm tree. Eventually somebody in a boat crossed the lake and rescued the elderly yet very fit tsunami survivor.

The 1946 tsunami led to efforts to predict future events. Today the Pacific Tsunami Warning Center (PTWC) provides bulletins for Hawaii and international countries, including the Indian Ocean. The West Coast and Alaska Tsunami Warning Center covers most of the North American cost including Puerto Rico and the US Virgin Islands.

As the previous tales indicate, the usual culprit in tsunami formation is a vertical displacement of the sea floor caused by a very powerful earthquake often near the edges of tectonic plates. The "Ring of Fire" is located on the margins of the Pacific Plate. Other tsunami causes are underwater landslides, subsurface volcanic eruptions, and if you are truly unlucky, an extremely rare large meteorite impact.

Once a tsunami is generated, it can

cause major problems near the closest land, as in the 1946 earthquake. The remaining energy is transmitted as low amplitude (height) but very powerful waves, which travel at speeds of near 500 MPH in deep water. The interval or period between waves can range from a few minutes to possibly an hour. During a tsunami event several waves of varying heights are often produced, and dangerous conditions near an affected coast may persist several hours.

When the tsunami waves approach the coast they slow down and greatly increase in height. The water may be moving 20 to 40 MPH when it crosses the shoreline. Sometimes a tsunami is not very steep and appears like a very rapidly changing tide that far exceeds normal ranges. In other cases it appears as a foaming wall of water or breaking wave. The first indication of trouble could be an elevated horizon and a very deep roar much louder than conventional surf. Or if a trough arrives first, the ocean could drain well below normal sea level, exposing reefs and flapping fish. In either case you have but brief minutes to escape to high ground.

Modern tsunami research permits much more accurate forecasts, greatly reducing the false alarm rate. When seismographs detect a powerful earthquake with tsunami potential, the first job is to determine the depth. Earthquakes over 30 to 40 miles deep rarely displace the ocean surface significantly. Did the earthquake cause a vertical or horizontal displacement near the sea floor? Horizontal fault motion may generate significant earthquake damage but no tsunami.

To determine the actual wave height, tide gauges can measure long period changes in sea level that are equal to expected tsunamis wave periods. A new generation of DART buoys measure pressure changes within the ocean associated with both normal tides and tsunamis. Fortunately many tsunamis are only minor non destructive sea level changes, but tsunami height prediction techniques have plenty room for improvement.

My research indicates that the last major Pacific-wide tsunami was in 1964. By all historical measures we are way overdue for a destructive event. I hope we are ready when the time comes.

You may wonder what was the largest tsunami ever measured. Look no farther than Alaska to a partially enclosed deep fiord like bay backed by glaciers and steep slopes. One July 1958 evening a near magnitude 8.0 earthquake moved the nearby Fairweather Fault 21 feet horizontally and 3 feet vertically. A tremendous rockslide made a huge splash inside Lituya Bay clearing trees more than 1700 feet above sea level. Incredibly, two out of three fishing boats anchored a short distance away rode out waves perhaps briefly 50 to 100 feet high. The third vessel was smashed to pieces with no survivors.

METEOROLOGY AND STORM SPOTTERS

This concludes the meteorology chapter for storm spotters. I hope you have learned more concerning how dangerous and interesting weather and some geological hazards work. The information you learned will enhance the quality of your reports.

Many dedicated spotters become very good amateur meteorologists in their local area. Careful observations of wind speed and direction, temperature, humidity, pressure changes, and cloud formations all provide short term weather outlooks. Doppler radar and satellite imagery can greatly expand a spotter's view.

When online some spotters enjoy evaluating numerical weather forecasting models. The model physics is far beyond the scope of this chapter. But some model displays show surface and upper level weather patterns for more than a week ahead of time.

Every model has strengths and weaknesses. The numerical equations that models use must make many assumptions or approximations of the atmosphere. These inevitably lead to errors in forecast accuracy. In general, model (and weather forecast) accuracy decreases the farther out in time you project. An average of many model runs under different assumptions is called an *ensemble forecast*. Often this consensus is the most accurate averaged over time, but there will always be exceptions. Sometimes the weather is highly predictable, and most models give fairly consistent results on consecutive future runs. At other times, there are large differences of opinion that change or "flip flop" over time. This means changing weather patterns and low confidence in the model forecasts.

I believe weather forecasting is gradually improving for many reasons. But surprises are inevitable. There will always be a need for talented SKYWARN spotters and human forecasters.

REFERENCES
[1] "Rare Tornado Touches Down in Alaska", *USA Today*, August 2, 2005.
[2] A. Watson, T. Turnage, K. Gould, J. Stroupe, T. Lericos, H. Fuelberg, C. Paxton, J. Burks, "Utilizing the IFPS/GFE to Incorporate Mesoscale Climatologies Into the Forecast Routine at the Tallahassee NWS WFO," 2009.

Hurricanes

In 1772 two hurricanes struck the Caribbean island of St Croix. A young man working on the island as a clerk wrote to his father in the American colonies and described the hurricane: "Good God! What horror and destruction! It is impossible for me to describe it or for you to form any idea of it . . . A great part of the buildings throughout the island are leveled to the ground; almost all the rest very much shattered, several persons killed and numbers utterly ruined — whole families roaming about the streets, unknowing where to find a place of shelter. . . In a word, misery, in its most hideous shapes, spread over the whole face of the country."[1]

The young man had no other family on the island and little reason to stay there except that he had no means to leave. Local residents took up a collection so that he could get an education and make a better life. The money was raised and he went to King's College (now Columbia University). In 1789 he was appointed by George Washington to be the nation's first Secretary of the Treasury: Alexander Hamilton.

Hamilton's description of the 1772 hurricane would be repeated by different people, with slightly different words, many more times over the next 200-plus years: Galveston, Texas in 1900, Okeechobee Hurricane in 1928, New England in 1938, Audrey in 1957, Camille in 1969, Andrew in 1992, and Katrina in 2005.

Of all the severe weather events that the United States faces each year, few can compare to hurricanes and tropical storms. Unlike other weather events that may impact a fairly localized area, hurricanes impact large areas covering not only states but entire countries. Hurricane Katrina, for example, impacted the Bahamas, Cuba, and the northern Gulf of Mexico coast. It then made its way inland dumping rain as far north as Quebec. Hurricane Ike packed hurricane force winds well into the Ohio River Valley.[2]

Amateur radio operators who respond to hurricanes do so as emergency and disaster communicators as well as storm spotters. In this section we will treat the amateur radio response to hurricanes as a combination of Amateur Radio Emergency Service (ARES) response and storm spotter response. We will look at the mechanics of a hurricane, related weather events, some history behind the amateur radio response to hurricanes, and storm spotter reporting criteria. Later we will look at real world experiences from several different amateur radio perspectives — a resident of an area impacted by hurricanes, an amateur radio operator sent into an area to assist with communications during and after a hurricane, and ARRL Section Managers from Hurricane Katrina and Hurricane Sandy.

Let's start with some basic meteorological information about hurricanes and tropical storms. The following introduction to hurricanes is written by Victor Morris, AH6WX.

AN INTRODUCTION TO HURRICANES

Hurricanes and tropical storms have long been feared as one of nature's worst hazards. Columbus encountered them in his voyages near the Caribbean. In 1635 a very powerful hurricane brought great destruction to early Pilgrim settlers in Massachusetts. Hundreds of these storms have seriously impacted the United States in recent centuries.

A hurricane is an intense low pressure area that forms over tropical or sub-tropical ocean waters. It is accompanied by a counterclockwise (in the Northern Hemisphere) rotating spiral band of heavy showers and thunderstorms. The thunderstorms may exceed 50,000 feet high, and are capable of producing torrential rain. The winds are relatively light at the outskirts of the storm, but increase

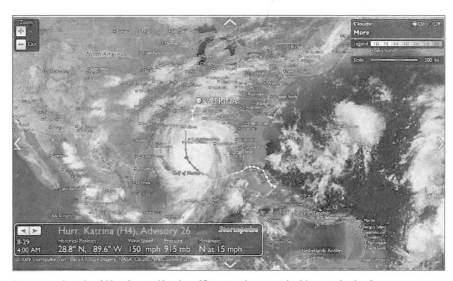

Image and path of Hurricane Katrina. [Stormpulse, used with permission]

The Saffir-Simpson Hurricane Scale

The Saffir-Simpson Hurricane Wind Scale is a 1 to 5 rating based on a hurricane's sustained wind speed. This scale estimates potential property damage. Hurricanes reaching Category 3 and higher are considered major hurricanes because of their potential for significant loss of life and damage. Category 1 and 2 storms are still dangerous, however, and require preventative measures. |In the western North Pacific, the term "super typhoon" is used for tropical cyclones with sustained winds exceeding 150 MPH.

Category	Sustained Winds	Types of Damage Due to Hurricane Winds
1	74-95 MPH 64-82 kt 119-153 km/h	**Very dangerous winds will produce some damage:** Well-constructed frame homes could have damage to roof, shingles, vinyl siding and gutters. Large branches of trees will snap and shallowly rooted trees may be toppled. Extensive damage to power lines and poles likely will result in power outages that could last a few to several days.
2	96-110 MPH 83-95 kt 154-177 km/h	**Extremely dangerous winds will cause extensive damage:** Well-constructed frame homes could sustain major roof and siding damage. Many shallowly rooted trees will be snapped or uprooted and block numerous roads. Near-total power loss is expected with outages that could last from several days to weeks.
3 (major)	111-129 MPH 96-112 kt 178-208 km/h	**Devastating damage will occur:** Well-built framed homes may incur major damage or removal of roof decking and gable ends. Many trees will be snapped or uprooted, blocking numerous roads. Electricity and water will be unavailable for several days to weeks after the storm passes.
4 (major)	130-156 MPH 113-136 kt 209-251 km/h	**Catastrophic damage will occur:** Well-built framed homes can sustain severe damage with loss of most of the roof structure and/or some exterior walls. Most trees will be snapped or uprooted and power poles downed. Fallen trees and power poles will isolate residential areas. Power outages will last weeks to possibly months. Most of the area will be uninhabitable for weeks or months.
5 (major)	157 MPH or higher 137 kt or higher 252 km/h or higher	**Catastrophic damage will occur:** A high percentage of framed homes will be destroyed, with total roof failure and wall collapse. Fallen trees and power poles will isolate residential areas. Power outages will last for weeks to possibly months. Most of the area will be uninhabitable for weeks or months.

Source – National Hurricane Center

Visible and infrared images of hurricane Joaquin (2015) from NOAA-19 [Courtesy Joe Carcia, NJ1Q]

Hurricane Joaquin making landfall on Turks and Caicos. [John Lawson, photo]

Volunteers Bill Kollenbaum, K4XS (left), and Gene Roll, KM4FUD (center), meet with Puerto Rico Section Manager Oscar Resto, KP4RF (right), at Red Cross HQ San Juan to discuss response efforts to hurricane Maria. [Mike Corey, KI1U, photo]

dramatically as one approaches the central storm core or eye. The worst winds are located in the eye wall, a ring of deep convective clouds that surround a much calmer center. An eye often begins to appear on satellite imagery when sustained (one minute average) winds reach 75 to 85 MPH.

It is useful to define a few terms frequently used in National Hurricane Center (NHC) advisories.[3] A *tropical disturbance* is a disorganized area of bad weather that lacks a well-defined low-pressure center. A *tropical depression* has definite rotation around a circular low-pressure area and maximum winds less than 39 MPH. The next step up is a *tropical storm*, with wind speeds from 39 to 73 MPH. *Hurricane force* is 74 MPH or higher. In the most extreme hurricanes sustained winds can be near 175 MPH. Note that tropical depressions, tropical storms, and hurricanes are all known as tropical cyclones.

The media often focus much attention on the peak winds of a tropical cyclone. This is certainly critical information. But a second factor, often overlooked, is the storm diameter. There can be huge variations in the area covered by damaging winds. Charley (2004) had a 25 mile radius of hurricane force winds while approaching Punta Gorda, Florida. Katrina (2005) and Ike (2008) brought hurricane force winds up to 125 miles from the center in their northeast quadrants.

Many factors need to come together to cause a hurricane to form. There must be an area of disturbed weather located over warm waters, generally 80 °F or warmer. The atmosphere needs to be moist and unstable, allowing deep convective clouds to form and persist. There should be very little wind shear (changes in wind speed and direction) near the disturbance from the surface to 10 miles up. Wind shear tears the tops off thunderstorms, inhibiting the development of many candidate bad weather areas.

The official hurricane season is June 1 to November 30 in the Atlantic Basin, and May 15 to November 30 in the Eastern North Pacific.[4] There are rare occasions when tropical cyclones form out of season. In the Atlantic Basin the long term average season brings 10 named storms. Names are given when a tropical cyclone attains tropical storm intensity. Typically six of the storms will become hurricanes, and two or three will reach major hurricane status (peak sustained winds 111 MPH or higher). But in the past century seasonal storm totals have ranged from one (1914), up to 28 (2005).

In the tropical Eastern North Pacific area (Central America to 140° west longitude), an average season brings 16 named storms and nine hurricanes. The Central North Pacific region (140° west to the International Date Line) averages three named storms per season and one or two hurricanes. These ocean regions also have large fluctuations in storm totals from year to year. When the tropical eastern Pacific becomes warmer than usual (the El Nino phenomenon), storm totals are often above normal. Atlantic tropical cyclone tracks and frequencies vary considerably as the hurricane season passes. June storms tend to form over the western Caribbean and southern Gulf of Mexico. Although a named storm is likely only once every other year on average, some of the early storms reach land. July storms follow much of the June pattern, but the possibility of tropical cyclone formation east of the Caribbean rises as the month passes. During August hurricane formation increases rapidly, reaching a peak near September 10. Almost any warm water region from the coast of Africa to the Americas can have tropical cyclone formation. These two months bring the highest risk of United States East and Gulf coast direct hits. By October wind shear increases in the eastern tropical Atlantic and few storms form there. The majority of October systems form along a broad band extending from the NW Caribbean Sea to the SW North Atlantic. November infrequently has significant hurricane development. Most very late season systems form near the western Caribbean.

But almost every good rule has an exception. In 1985 Hurricane Kate struck the Florida Panhandle during Thanksgiving week.

Every coastal state from Maine to Texas has experienced strikes from full hurricanes. The maximum risk areas based on history are South Florida, the Outer Banks of North Carolina, and the parishes of southeastern Louisiana. Elevated risk regions include most of the remaining Gulf Coast and the Carolinas. Medium risk

Tropical Cyclone-Induced Tornadoes

Tropical cyclones such as hurricanes and typhoons are responsible for a wide range of related weather phenomena, including flooding, wind damage, storm surge, and tornadoes. Tornadoes pose a significant risk to life and property whether they are formed by severe thunderstorms or tropical cyclones. Tropical cyclone-induced tornadoes differ from severe thunderstorm-induced tornadoes in several ways. Typically they are shorter lived and somewhat weaker than those that occur in the Great Plains. Tropical cyclones also rely on warm tropical air as their energy source. As such, tropical cyclone-induced tornadoes are not usually accompanied by hail or a lot of lightning.

Tropical cyclone-induced tornadoes are most frequently produced in the right-front quadrant of the tropical cyclone in the spiral rain bands within 50 – 200 miles of the center of the storm. These rain bands can develop into severe thunderstorms that can produce tornadoes.

If we look at these tornadoes from the storm spotter's perspective, we can see how they can pose serious threats. The storm spotter's role is to provide ground truth reports to aid in the issuance of warnings. During a tropical cyclone it may be difficult or even impossible for a storm spotter to observe anything more than what they can see from their home or other secure location. The ability to see a tornado is often hindered by the tornado being rain wrapped. Since these tornadoes are often of a shorter duration than conventional tornadoes a trained spotter may not be there to observe the event. Amateur Radio operators are likely to be involved in an Emergency Operations Center (EOC) or other emergency communications operations. So where does the NWS get information on tornadoes during a tropical cyclone? Doppler radar is the primary source of information. Notification may also come in from a variety of sources but trained spotters are usually limited. Public safety officials, the news media and the general public also provide reports. Overall, tropical cyclone-induced tornado ground truth is harder to come by. This combined with the aforementioned challenges of tropical cyclone-induced tornadoes typically results in a higher warning false alarm rate.

Hurricane Ivan, which struck the Florida panhandle and Alabama in 2004, provides us with an extreme example of a tropical cyclone-induced tornado outbreak. Ivan resulted in eight deaths and 17 injuries from tornadoes alone. A total of 117 tornadoes occurred over a three-day period. The tornado damage ranged from F0 to F2. Most of the tornadoes took short tracks ranging from a few miles to about 20 miles. The Florida Panhandle experienced some of the most significant impacts from tornadoes associated with Ivan. Between September 15 and 16 the NWS Tallahassee office received 24 tornado reports and issued 130 tornado warnings, roughly a 10% higher warning false alarm rate as compared to the 2004 national average.

Amateur Radio storm spotters must take tropical cyclone-induced tornadoes seriously. Despite being weaker and shorter-lived than conventional tornadoes, they still pose a significant threat to life and property. Accurate ground truth reports are critical to the warning system for this hazard. But ground truth reports are never more important than spotter safety. When spotting from an area impacted by a tropical cyclone you should make sure you follow all safety precautions. Do not go out spotting mobile when road or weather conditions make it unsafe to do so. If you are operating from an EOC or other secure location you may end up receiving weather reports to relay to local emergency management, public safety, and the NWS. Use your knowledge as a trained storm spotter to relay valuable weather information.

Damage from a tornado caused by Hurricane Ivan. (Dave Barton, AI4GF, photo)

areas are northeastern Florida, Georgia, Virginia, Long Island, and near Cape Cod. All other Atlantic coastal locations have a slightly lower risk, but any one year may still bring a devastating hurricane.

Out in the eastern and central North Pacific, the majority of tropical cyclones move west to northwest throughout the season which peaks in August. Some storms may turn northeast late in the hurricane season. Cool ocean currents limit the area of 80 °F or warmer water. That means not too many systems survive to get far beyond the tropics. Western Mexico is an occasional target of mostly mid to late season hurricanes, and a few have struck the Hawaiian Islands. The worst storm, Hurricane Iniki brought sustained winds near 140 MPH to Kauai in 1992. Only one minimal hurricane in 1858 held together long enough to reach San Diego, California.

Tropical cyclones move along with the atmospheric winds surrounding them; most of the time these currents carry them westward in lower latitudes, although there are exceptions. As a hurricane gains latitude, it will generally encounter the middle latitude westerlies. This will cause many tracks to bend to the north then northeast, a process known as recurvature. The process of identifying the location of recurvature is a very challenging forecasting issue. Storms that run out of well defined steering currents may stall, loop, or do other erratic things. This makes it very difficult for hurricane forecasters to issue accurate warnings, which becomes critical if the system is near land. Once recurvature is completed a tropical cyclone often accelerates to forward speeds of 30 – 50 MPH. The rapidly approaching storm may permit only short periods of accurate warnings for anyone in harm's way.

The hurricane life cycle can vary as much as that of a human being. Some form quickly and rapidly fall apart. Others may go through several periods of weakening and strengthening. A typical tropical cyclone lasts a few days to slightly over a week. The records range from a few hours to an extreme longevity case of 31 days. This record was held by Pacific Hurricane John in 1994. He initially formed off Mexico, passed south of Hawaii and continued into the Western North Pacific, reclassified as a typhoon. Later he curved, crossed the International Date Line, and became a hurricane again. John finally lost tropical features as he neared Alaska.

Most hurricanes finally die off for one of three reasons. If they move inland, the storm loses its vital oceanic heat and moisture source and gradual weakening occurs. The second is motion over higher latitude waters too cold to sustain concentrated tropical convection. But mariners need to recognize that some tropical cyclones can transition into dangerous extratropical storms. The third reason a tropical cyclone can weaken is increasing wind shear often combined with the intrusion of dry, stable air into the storm circulation.

Now is a good time to examine damage and safety issues caused by tropical cyclones. Most people think of the wind first when they hear the word hurricane. The wind force exerted perpendicular to an exposed surface is proportional to the wind velocity squared. There is some debate among structural engineers regarding the wind stress calculations. A simple estimate used in the past is:

$$F = 0.004 \times W^2$$

where F is force in pounds per square foot and W is wind speed in MPH.

Using this approximation, 50 MPH winds create 10 pounds per square foot of force. Most people find walking difficult at this wind speed. 100 MPH wind brings 40 pounds per square foot, and 150 MPH creates 90 pounds per square foot of force.

Not only is the maximum sustained wind force an issue. Peak hurricane gusts may exceed the one-minute sustained wind by 20 – 50%. The gusts are enhanced by turbulence generated as the winds cross rough and irregular natural terrain, buildings, and other manmade structures. These features are also likely to create sudden changes in wind direction. Very rapid changes in force will attack exposed structures, potentially for many hours. I hope your home is prepared for this!

Amateur radio operators have a special concern during hurricane force winds. Every amateur with a tower and/or Yagi or similar antenna should know the engineered wind rating for his setup. There is a risk stronger hurricane forces will exceed design parameters. In such cases one should plan to safely lower towers and take down larger square footage antennas. Complete all work at least 24 hours ahead of the first predicted strong winds.

Even vertical and wire antennas may be at risk depending on wind exposure. Any antenna can be instantly destroyed by flying debris which is not counted in wind force calculations. I suggest having a sufficient supply of simple dipole wire antennas for post-storm HF operations, plus appropriate verticals for VHF/UHF.

Tornadoes are often a surprise consequence of a tropical cyclone passage. More than 100 separate tornadoes have accompanied a few hurricane landfalls. They typically form under spiral band lines of thunderstorms. Frequently the tornadoes will develop in the right semicircle of the storm 50 – 300 miles from the center, but other locations are very possible. The twisters in tropical systems often form and dissipate rapidly. This allows local National Weather Service (NWS) offices only a very short time to issue to issue tornado warnings. SKYWARN spotter reports are critically important for these events.

Water is a major cause of tropical cyclone damage. A cubic foot weighs almost 64 pounds, and that can generate unbelievable forces when it starts moving. There are several water related tropical cyclone dangers.

Flat terrain locations can accumulate significantly large and deep areas of standing fresh water. Many tropical cyclones including depressions will drop several inches of rain per 24 hours. Wetter systems have dropped a foot to more than three feet of rain in a day. This can happen hundreds of miles from the coast. In hilly locations the runoff is much faster and sudden flash flooding becomes a major concern. Flash floods can be extremely dangerous when high intensity precipitation cells repeatedly "train" over the same location. When you are driving, remember the mantra "turn around, don't drown." Flowing water only two feet deep can sweep many autos away.

If heavy tropical rains move over a saturated river drainage basin, general major river flooding can be expected. This type of flooding is more predictable, hopefully allowing enough time for evacuations and mitigation measures such as sand bags.

Coastal low lying beach communities face the life threatening dangers of storm surge and wave battering. A storm surge is a rise of sea level higher than the astronomical tide. It is created primarily by the frictional stresses of violent onshore wind, primarily in the right semicircle of the hurricane at landfall. Many factors contribute to surge height including peak winds, the area covered by very high wind, and the amount of time the winds blow. The underwater terrain or bathymetry can play a significant role. All things being equal, locations with shallow bottoms well offshore have higher surge but smaller wind waves. Areas with a narrow continental shelf will have larger waves but a lower maximum surge.

It is worth noting how much difference the hurricane diameter can make on the observed storm surge. Hurricane Charley, with a 25 mile radius and 145 MPH peak, generated only a 6 to 8 foot surge in southwestern Florida. Hurricane Katrina, 120 MPH at final landfall in Mississippi and a 125 mile radius, generated an all-time United States surge record. NHC investigators confirmed a 28 foot surge at Pass Christian, Mississippi.

There has been a slow but consistent improvement in NHC track forecast accuracy in recent decades. This has happened due to the mix of vastly increasing computational power, improving models of hurricane physics, and better multi-platform observational networks. The latter includes increasing varieties of weather satellite data, better instrumentation on hurricane hunter aircraft, buoy and volunteer ship observations, and many more real time reports from observers like you the SKYWARN spotter. At the same time in our high-tech age, a dedicated and experienced hurricane forecaster can still improve significantly on forecasts generated from computer model output.

Meanwhile there has been somewhat less success in reducing intensity change forecasting error. Hurricanes that intensify rapidly are especially difficult to identify precisely ahead of time. We know the general factors that promote rapid intensification, but there is still considerable difficulty in applying the factors to specific hurricanes. This situation can be very dangerous should a hurricane undergo rapid unexpected strengthening in the last few hours prior to landfall.

Finally, the NHC forecasters are well aware of Murphy's Law when they write their forecasts. There are unknowns that will create forecast errors. In order to play it safe, tropical storm and hurricane warnings are issued for a considerably wider area than the regions which actually experience damaging winds. The goal of the forecasters is that 95% of the time (two standard deviations for the statistically inclined) the dangerous wind speeds will occur somewhere within the warned areas. But that safety factor means that in quite a few cases you will experience conditions less severe than predicted. As technology continues to improve, it will enable forecasters to specify tropical cyclone danger areas more precisely in the future.

REPORTABLE CRITERIA

During a hurricane or tropical storm, SKYWARN activation, as well as activation of the NHC station WX4NHC and the Hurricane Watch Net, is essential in getting ground truth reports to the NWS/NHC. Let's look at a couple different ways these reports are received by the NWS.

First are the SKYWARN storm spotters. Spotters in the area where the hurricane makes landfall can relay important information about flooding due to intense rain or storm surge. Rainfall amount can also be relayed. Also critical is the time the rain began and ended. This can give a good picture on hourly rainfall totals for a particular area. Damaging wind, another characteristic of hurricanes, is also important information to relay. Wind speeds, both sustained and gusts, are important information. It is possible to estimate wind speeds and these estimates can also be reported. Also report associated damage with the winds, for example "roof from house blown off" or "pine trees, approximately 75 feet tall uprooted." Damage from wind can help meteorologists determine wind speed. Remember that hurricane force winds are in excess of 74 MPH. Winds at this speed will turn many things into flying missiles! Stay inside in a safe location. *Do not* go outside to try and observe weather conditions when it is this dangerous.

It is also important to relay tornado activity to the NWS. As hurricanes make landfall tornadoes can be generated in the rain bands, particularly in the upper right quadrant of the storm. Getting accurate information about these tornadoes is important in issuing warnings for areas that may be impacted by them. But again, it absolutely critical to stay safe! Do not go out storm spotting if it means compromising your safety. And remember, tornadoes caused by hurricanes can appear with little or no warning. While they are typically weaker and of shorter duration they are no less deadly.

The National Hurricane Center in Miami, Florida, is home to WX4NHC which works in conjunction with the Hurricane Watch Net. This net provides a vital service in relaying hurricane information to the NHC and NWS. Report information that is needed is similar to reportable criteria for SKYWARN, but with a few minor differences. The reporting station's location, along with wind speeds, both sustained and gust, are helpful. Again these may be measured or estimated. Wind direction is also needed. Barometric pressure, in inches or millibars, is also reportable along with sea state, precipitation reports, and, where appropriate, damage reports.

Whether you are relaying reports via a SKYWARN net or through the Hurricane Watch Net, don't forget the critical information: who, when, where, and how to contact. Include your call sign or name in your report. Include the means to contact you, such as email, cell phone, or IM. Include your location as a physical address, latitude / longitude, or proximity to a known location, for example, "I'm eight miles north of Biloxi, Mississippi on highway 15." When giving the date and time make certain you are giving the date and time that the weather event was observed. There are times where the relay of a report may be delayed.

Finally don't forget post-storm reporting. Information on storm damage is important to the NWS. If you have pictures of storm damage send them to your local NWS WFO. Many WFOs have information on submitting pictures or video on their websites.

WX4NHC — Amateur Radio at the National Hurricane Center

By Julio Ripoll, WD4R

WX4NHC is the amateur radio station located at the National Hurricane Center in Miami, Florida on the campus of Florida International University.[5] The station has been totally assembled from donated equipment and has been a permanent public service to the Hurricane Center since 1980.

The WX4NHC Team is composed of 20 specially trained volunteer operators who man the station in 3 hour shifts. For example, during the historic 2005 hurricane season the station was manned, sometimes with two to three operators at a time, for more than 500 hours. WX4NHC has operated twice from inside the eye of a hurricane (Katrina and Wilma) and collects hundreds of reports each hurricane season. During hurricane Dorian in 2019, we maintained direct communications with stations in the Bahamas during approach and landfall and relayed real-time surface reports received to National Hurricane Center Hurricane Specialists.

WX4NHC is located at the National Hurricane Center in Miami, Florida. [Julio Ripoll, WD4R, photo]

Kenneth Graham, WX4KEG, Director of the National Hurricane Center (seated) with NHC Amateur Radio Coordinators John McHugh, K4AG, and Julio Ripoll, WD4R. [Julio Ripoll, WD4R, photo]

The WX4NHC team has been nationally recognized for their volunteer international humanitarian efforts during hurricanes and also for their Haiti Earthquake mission in 2010 where WX4NHC/HH2 setup a HF and VHF station that operated inside the University of Miami's Field Hospital in Port-au-Prince for five weeks. This station became the hub for emergency communications between the UM Field Hospital, US Naval Hospital Ship *Comfort* and local NGOs and the UM hospital doctors in Miami.

WX4NHC Purpose and Goals

• Collect weather data "Surface Reports" from hurricane-affected areas in real time for use by the Hurricane Specialist at the National Hurricane Center.

• Provide backup emergency communications to and from the Hurricane Center over amateur radio during and after a direct landfall in South Florida.

• Provide Hurricane Advisories over amateur radio, when other sources are not available to the affected areas.

• Enhance and promote the accuracy and availability of measured weather data surface reports.

• Our Mission: To help save lives.

Activation of WX4NHC

WX4NHC is activated whenever a hurricane is within 300 miles of landfall in the Atlantic, Caribbean, Gulf or the eastern Pacific oceans. Emergency backup communications are also provided from WX4NHC for communications between NHC and NWS offices and other local and national government agencies.

The "Surface Reports" sent in by stations in the affected area provide the forecasters with supplemental weather and damage data that are not normally available to them and are frequently incorporated into their advisories as they provide a human perspective and eye witness accounts of what people are experiencing during a hurricane.

Communications Modes Used by WX4NHC

WX4NHC uses many modes to communicate and collect Surface Reports from stations in the affected areas.

• *HF*: Hurricane Watch Net 14.325 MHz primary frequency, 7.268 MHz secondary frequency

• *VHF/UHF*: for local communications as well as SARnet statewide linked repeater system.

• *APRS*: HF (30 meters LSB 10.151 MHz) and VHF (144.390 MHz)

• *VoIP*: VoIP Hurricane Net EchoLink Conf Room WX+Talk node 7203, IRLP node 9219

• *Winlink*: wx4nhc@winlink.org (email subject line must start with "//WK2K".

• *D-Star/D-Rats*: voice net RF002A, Net Control will relay reports via D-Rats report form.

• *CWOP*: Citizen Weather Observer Program: Hams can send their weather instrument data automatically via APRS and non-hams using the internet to NOAA's Mesonet Map website.

• *ON-NHC*: Observers Network for National Hurricane Center: Hams and non-Hams can send their hurricane surface reports using online form at **www.wx4nhc.org**.

THE HURRICANE WATCH NET

By Bobby Graves, KB5HAV

Hurricanes have been a part of nature since the beginning of time. History records the tracking of these storms, as well as forecasting tracks and issuing warnings, as early as 1870.[6]

In 1965, a time in which hurricane forecasting was still in its infancy, Hurricane Betsy came roaring across the Bahamas, the Florida Keys, and then hammered southeast Louisiana. Betsy was the first hurricane to make landfall in the United States that resulted in over $1 billion in damages.[7] The storm became known as "Billion Dollar Betsy."

While stationed at the US Naval Mobile Construction Battalion Center in Davisville, Rhode Island, Jerry Murphy, K8YUW, handled countless phone patches and messages to and from military deployed personnel as a member of the Intercontinental Amateur Radio Net (IARN) on 14.320 MHz. As Hurricane Betsy was moving through the Bahamas, Jerry, along with Marcy Rice, KZ5MM located the Panama Canal Zone Panama, helped relay weather information between those in south Florida and the Bahamas. There was so much interest in what the

March 24th, 2018

A note to the American Radio Relay League and all volunteer amateur radio operators around the world,

Greetings my fellow HAM radio operators. As the incoming Director of the National Hurricane Center, I want to take a moment to thank you for the important work you do to ensure communications continue during and after a disaster.

In the aftermath of Hurricane Katrina, HAM radio was the only remaining method of communication for many of us in the disaster zone. My handheld and the base unit at the New Orleans National Weather Service Office were our only links to the outside world. I cannot stress enough how critical redundant methods of communications are during and after a disaster. As the infrastructure fails, it is you all that keep the two-way dissemination of critical information going.

We at the National Hurricane Center recognize and value the role of WX4NHC and all of the amateur radio operators with whom they connect. As we saw during last year's extremely busy hurricane season, the next hurricane to impact our area is always just around the corner. After experiencing the aftermath of Hurricane Katrina, I can tell you we can't do without WX4NHC to ensure we can continue to perform our mission responsibilities.

We too have to be ready for the next storm. Testing WX4NHC is vital to ensure we are ready for anything the hurricane season will bring.

I cannot thank all the amateur radio volunteers enough for their ongoing dedication to our shared mission of saving lives. 73's to you all and I look forward to working with my fellow HAMs.

Kenneth E. Graham, WX4KEG

National Hurricane Center

Kenneth Graham, WX4KEG, will take over leadership of the National Hurricane Center in Miami, Florida, on April 1.
(NOAA photo)

storm was doing that it created a major disruption in IARN activities. Jerry suggested to the net manager to move those interested in the storm up 5 kHz to get them off the net, and the net manager agreed. Marcy followed Jerry, and together they established the first Hurricane Watch Net on 14.325 MHz.[8]

In a letter dated March 1999 from Jerry Murphy, he recalls the following: "We received the latest advisories and bulletins from various Florida hams; one of them worked for the city of Miami. Sometime later, maybe a year or two later, Ellie Horner, K4RHL subscribed to a teletype network and had a RTTY system installed in her home. That kept us better informed in future storms."

Fifteen years later, in 1980, the National Hurricane Center saw its first amateur radio station installation using the call sign W4EHW. When W4EHW (operating as WX4NHC as of 2004) opened for business at the NHC, it became a lot easier for the HWN member stations to provide real-time information to those in the areas affected by tropical storms. Ground truth observations were gathered and relayed to the NHC's forecasters in a much timelier manner.

In the decades since Jerry Murphy's efforts in "Billion Dollar Betsy," there have been more than 550 named storms in the Tropical Atlantic basin. More than 130 storms have come ashore as a Category 1 hurricane, or higher, and the Hurricane Watch Net was active for each one. The net also activated for many tropical storms as they were either forecast to become a hurricane prior to landfall or there was a request by forecasters at the National Hurricane Center to do so.

The Hurricane Watch Net generally activates whenever a hurricane is within 300 statute miles of expected landfall. Our area of coverage includes the Caribbean, Central America, eastern Mexico, eastern Canada, as well as all US coastal states.

When activated, you will find us on 14.325 MHz by day and 7.268 MHz by night. Why do we state these frequencies without a plus or minus amount? Because those who are operating using marine radios have to program in the frequencies — marine radios do not have a VFO or RIT. Furthermore, these two frequencies come preprogrammed into many marine radios. Many non-hams listen in via shortwave radio and know this is where to find us when are activated. Before any net activation, if either frequency is in use, we always ask permission to use them. Additionally, it is our practice of being on the air ahead of the amateur radio station at the National Hurricane Center — WX4NHC — to establish our net operating frequency, to issue advisory data, and to line up reporting stations. It helps us tremendously to know the operators' locations, names, and weather measuring capabilities in advance of the storm's arrival.

There are three primary purposes of the HWN:

1) Disseminate the latest advisories issued by the National Hurricane Center. We do so for marine interests, Caribbean Island and Central American nations, and other interests where public media is not readily available. During hurricane events, these people receive their weather information from the United States, mostly by amateur radio!

2) To obtain real-time ground level weather conditions and initial damage assessments, from amateur radio operators in the affected area and relay that information to the forecasters at the National Hurricane Center by way of WX4NHC.

3) To function as a backup communications link for the National Hurricane Center, National Weather Service Forecast Offices, the Canadian Hurricane Centre, Emergency Operations Centers, Emergency Management Agencies, Non-Governmental Organizations, and other vital interests. This can involve military relief operations in the protection of life and property before, during, and after a hurricane event.

To achieve mission goals, the HWN relies upon its members…experienced net control operators. Some are seasoned ex-military and/or MARS operators. Others have gained their experience through public service roles. On average, there are 40 active members strategically located throughout the US, Canada, Mexico, Central America, and the Caribbean. To better assist those who speak Spanish only, HWN has a number of bilingual operators.

Every veteran HWN member has his or her own poignant memories of different storms. HWN stalwart Bob Botik, K5SIV (SK), particularly remembered Hurricane Mitch, the killer storm in 1999 that took more than 11,000 lives in Honduras alone (thousands are still missing and the final death toll may never be known) as it tore through Central America.[9]

"I remember the feeling of helplessness," said Botik, "as we listened to reports of flooding, the casualties, and, the damage to the homes of some of the poorest people on the planet. . ." Others recall communicating with the flight crews of the Hurricane Hunter aircraft as they flew into storms, or communicating with a British frigate operating in the Caribbean, or maintaining communications with a ham in an upstairs room of his home, operating with a wire antenna and a battery, as he watched while his neighbors' homes were ripped apart by the fury of the storm.

In 1999, the HWN was awarded the Outstanding Achievement Award by the National Hurricane Conference for continuing efforts to serve the public. In 2000, the ARRL honored the HWN with the League's International Humanitarian Award.

In one of his letters, Murphy explained the nature of the services provided by the HWN: "The guiding principles of the Hurricane Watch Net were to serve the Public Interest, Convenience or Necessity. Nothing more, nothing less." So it is today, more than 50 years later.

VoIP (Voice over Internet Protocol) Hurricane Net

The VoIP (Voice over Internet Protocol) Hurricane Net is a network that utilizes the Amateur Radio VoIP modes of EchoLink and IRLP (the Internet Radio Linking Project) to put together one large network of VHF/UHF stations to provide meteorological and damage reports to the National Hurricane Center and other agencies for the protection of life and property. The network started in 2003, as two separate nets, with one net on EchoLink on the *WX_TALK* EchoLink conference node: 7203 founded by Kevin Anderson, KD5WX. The other net was on IRLP Reflector Channel 9219, known as the Raleigh IRLP Reflector, and founded by Danny Musten, KD4RAA, with initial net management duties done by the late Robert Broderick, WE4B.

In the spring and early summer of 2004, the EchoLink and IRLP systems were merged into one network and called the VoIP Hurricane Net cofounded by Kevin, KD5WX, and Danny, KD4RAA. Technical developments on EchoLink and IRLP allowed for the combination of the two systems to make one large combined network. Tony Langdon, VK3JED, supported the technical maintenance of the network and made possible the technical innovations to combine the two VoIP modes allowing the *WX_TALK* EchoLink conference node: 7203 and IRLP 9219 reflector channel to be connected together.

Just after the start of the 2004 Hurricane Season, Rob Macedo, KD1CY, joined the VoIP Hurricane Net management team as the VoIP Hurricane Net Control Scheduler for the weekly VoIP Hurricane Prep Net and Net activations. Rob would be named the Director of Net Operations during the spring of 2005 by Kevin and Danny.

The VoIP Hurricane Net has a weekly net in the Atlantic Hurricane Season months of June through November on the *WX_TALK* EchoLink conference node: 7203/IRLP 9219 system at 8 PM ET. The net meets monthly in the months of December through May at 8 PM ET with the December net one hour earlier at 7 PM EDT. This change is to accommodate SKYWARN Recognition Day (SRD) where the EchoLink/IRLP system is used to host NWS offices on an hourly schedule so recognition can be given to Amateur Radio SKYWARN spotters across the country who help support their local National Weather Service Offices. WX4NHC, the Amateur Radio station at the National Hurricane Center, is also on the network during SRD. The prep net hosts occasional training sessions on various forms of severe weather, net control operation over EchoLink and IRLP and also has a "question of the week" or "question of the month" for net participants to answer during the prep net.

During hurricanes, power and Internet infrastructure can be the first items lost. The VoIP Hurricane Net has demonstrated the ability to be able to gather reports from stations in the affected area even at the height of hurricane activity. It's also noted that the VoIP Hurricane Net is a complement to the Hurricane Watch Net and other Amateur Radio communications modes and does not replace any of the infrastructure in place today to support hurricanes. Over the years the VoIP Hurricane Net has been able to provide many critical reports during hurricanes through the ability to connect to local VHF/UHF repeaters/links and there are a number of examples of this over the years.

When the net combined to form one large network in 2004, the VoIP Hurricane Net provided critical reports during Hurricane Ivan as it went through the Windward and Leeward Islands. First hand reports were received from stations in the affected area of some of the hardest hit islands including Tobago, St Vincent, Grenada and Barbados. In that same year, Florida was hit by four landfalling hurricanes including Hurricane Charley, Frances, Ivan and Jeanne. The VoIP Hurricane Net was able to relay numerous reports from Amateur Radio operators and SKYWARN spotters across all those landfalling hurricanes as they hit Florida.

In July 2005, Emily was expected to reach hurricane strength as it crossed the Windward Islands but remained a tropical storm and initially appeared that Emily would not intensify into hurricane. Suddenly in the late evening of July 13, Emily intensified into a hurricane. The VoIP Hurricane Net was able to activate and obtained numerous reports from the island of Grenada of hurricane force winds, damage to trees, power lines and small cars. The roof of the hospital was blown off in Carriacou, Grenada, along with other damage and flooding in Trinidad and Tobago. This was accomplished by Amateur Radio operators on various islands within the Caribbean operating on 75 meter HF via the Caribbean Emergency Weather Net and then sending their reports to the VoIP Hurricane Net. It was during the night, and long-range HF communication on 20 meters was not possible. This provided significant reporting from that area as Emily moved through the islands. Also, in 2005, the VoIP Hurricane Net had numerous reports during Hurricane Katrina and especially Hurricane Rita when it made landfall in eastern Texas through several Amateur Radio operators linked in on VHF/UHF reporting from key locations such as hospitals. Flo Garneau, WM6V, provided numerous reports of damage in Livingston, Texas, including problems at the Livingston Texas Memorial Hospital and a local school shelter. She also provided measured wind reports.

In August 2008, the VoIP Hurricane Net started working with the Crescent City Amateur Radio Group (CCARG) and this allowed for increased reporting from southeast Louisiana and the New Orleans area as Hurricane Gustav moved through the area. CCARG became a participant in the VoIP Hurricane Net on a regular basis. During Hurricane Isaac in August 2012, CCARG was a very critical path in obtaining reports from the New Orleans area and much of Southeast Louisiana. While Isaac was a Category 1 hurricane, it stalled over southeast Louisiana causing significant wind damage and storm surge flooding in the region.

In late October 2010, the VoIP Hurricane Net was active for Hurricane Tomas as it affected St Lucia, Barbados, and St Vincent. Reports from Amateur Radio operators and the storm blogger website http://stormcarib.com were sent to the WX4NHC, the Amateur Radio station at the National Hurricane Center, and the reports were used in advisories issued by the National Hurricane Center.

On August 27 – 28, 2011, Hurricane Irene impacted the Carolinas, mid-Atlantic and Northeast US coast with widespread pockets of wind damage, storm surge flooding and significant river, stream and urban flooding from heavy rainfall. The VoIP Hurricane Net combined with the Southern New England regional SKYWARN Network run out of WX1BOX, the Amateur Radio Station at the National Weather Service in Taunton Massachusetts, formed one large network providing significant reports to WX4NHC, the Amateur Radio Station at the National Hurricane Center, and other regional National Weather Service forecast offices. This same network was combined again on October 29, 2012 as Hurricane Sandy significantly impacted New Jersey, New York City and Long Island with significant effects as far north as New England and as far south as the Chesapeake Bay and Delmarva region. The reports received by WX4NHC through the combined network were significant with reports utilized in National Hurricane Center advisories. A liaison to the Canadian Hurricane Centre was also present during portions of these hurricane net activations.

More information on the VoIP Hurricane Net can be found via their website, e-mail group and social media outlets:
- Website: **www.voipwx.net**
- Yahoo Groups: **http://groups.yahoo.com/voip-wxnet**
- Facebook: **www.facebook.com/voipwxnet**
- Twitter: **www.twitter.com/voipwxnet**

THE HURRICANE EXPERIENCE

Thousands of amateur radio operators have had direct or indirect experience with hurricanes. In recent years several hurricanes have seen enormous responses from the amateur radio community. The stories amateurs can tell of their experiences, both good and bad, could fill volumes. For our purposes in this book it is worth looking at the experiences of amateur radio operators, each representing a different perspective. We will start with Vic Morris, AH6WX, who experienced eight different hurricanes over the course of six years while residing in Puerto Rico. We will then look at the experience of an amateur radio operator sent to provide emergency communications assistance during Hurricane Gustav. And finally we will look at Hurricane Katrina and Hurricane Sandy from the perspective of the Mississippi Section Manager Malcolm Keown, W5XX and the New York-Long Island Section Manager Jim Mezey, W2KFV.

Life in the Casa de los Hurricanes

(The Home of the Hurricanes)

By Vic Morris, AH6WX

Puerto Rico, a United States territory for over a century, is known for lush tropical scenery, beautiful beaches, and pristine azure waters. The island also lies close to the path of Atlantic hurricanes.

I moved to Rincon, Puerto Rico on the island's northwest coast in June 1994. As a former meteorology professor I was very aware of Caribbean hurricane history. Long term records indicate that Puerto Rico is struck by sustained hurricane conditions approximately once every eight years. As an amateur radio operator, then KP4WN, DX was a primary interest. I wanted a near ocean location favorable for HF propagation. I picked a well built concrete house on a ridge 200 feet above sea level and 1/4 mile from the Mona Passage. I chose a three-element tribander bolted to a 20 foot mast above my roof. It could be easily lowered by two persons, an exercise in which I was to become very proficient in the following years.

Rural northwest Puerto Rico did not have the reliable infrastructure that most residents of the 50 states rely on. I lived near the end of a leaky public water supply line built in 1928. Nearly everyone in my neighborhood had water storage tanks as a back-up when the public supply failed. The electricity was also erratic at times, partly due to 160 thunderstorm days per year. Many homes and businesses installed their own power generators. The need for self reliance in Rincon proved to be essential in many hurricane seasons.

The period of 1970 – 1994 was one of below average Atlantic hurricane activity. However, there were some very destructive storms such as Andrew (1992) and Hugo (1989). My first hurricane season in Puerto Rico proved to be tranquil with just one minor tropical storm encounter. But starting in 1995 a new cycle of active Atlantic seasons began which many meteorologists feel is still in progress. Sometimes for reasons not well understood, several consecutive hurricane seasons will bring storms that follow very similar tracks. This is what happened to the northeast Caribbean island region from 1995 through the year 2000.

Everything changed in 1995. The Atlantic produced 19 named storms, the most in decades. Some of these were the dreaded Cape Verde hurricanes that are spawned off the coast of Africa. Such systems frequently become major hurricanes, taking long lived west to west-northwest tracks that may threaten the Caribbean or the Americas.

Early in September large diameter Hurricane Luis was just under Category 4 status (131 – 155 MPH) while nearing Antigua and other northeastern Caribbean islands. Hurricane warnings were posted for many islands including Puerto Rico. I decided to take my beam down even though there were signs that Luis would turn northwest enough to avoid a direct hit on Puerto Rico. I experienced tropical storm force winds but no real damage. Just 10 days later Marilyn arrived on the scene from a somewhat lower latitude. I hadn't even gotten around to putting my beam back up! Marilyn, a smaller diameter hurricane, smashed into the US Virgin Islands with top sustained winds estimated at 110 MPH, just under Category 3. Rincon about 100 miles from the eye again received tropical storm force winds.

1996 was a year of unwelcome surprises. Mother Nature began to act up in early June as a minor tropical disturbance neared my QTH from the east. One late afternoon after work, I headed down to the beach to cool off. There were just a few cumulus clouds over the adjacent hills when I left my house. When I got down to the ocean a few minutes later, these clouds began to explosively build into thunderheads. Swimming during a thunderstorm

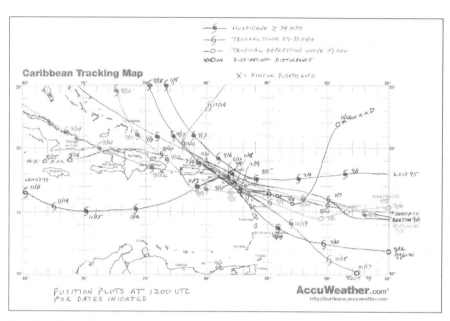

Plot showing paths of hurricanes through the Caribbean. [Courtesy Victor Morris, AH6WX]

is a bad idea. It is also not wise to leave sensitive electronics plugged into wall outlets most Puerto Rican rainy season (May to October) afternoons, due to the thunderstorm threat. Unfortunately I had done just that.

I quickly headed home, and started to rapidly unplug things. Suddenly a high amperage bolt hit my TH3JRS antenna. Some of the energy grounded properly, but some went through my guy wires into my roof. The remainder of the lightning flash arced over to a nearby power pole transformer. The electricity then entered my wiring, exploding one outlet and melting some wires. I was in my living room, and had just unplugged a TV when the lightning struck. It generated a blinding flash and deafening concussion much like a grenade explosion. If I had been a few seconds slower unplugging my TV, I would not be here writing this article. There was one bit of good news. My homeowner's policy and ARRL equipment insurance covered nearly all the damage. It took quite a while for me to get my radio gear repaired or replaced. Meanwhile the 1996 hurricane season ramped up into more serious action.

The majority of Cape Verde origin hurricane threats to Puerto Rico occur in August and September. But the 1996 season got a head start with the Independence Day formation of Bertha southwest of the Cape Verde Islands. The storm steadily developed, entering the northeastern Caribbean on July 7 at Category 1 force (74 – 95 MPH). Unofficial top gusts were just over 100 MPH in the US Virgin Islands. For the third consecutive threat, Rincon experienced tropical storm force winds, although there were reports of nearby waterspouts.

Hortense, the eighth named storm of 1996, approached the island of Guadeloupe with near hurricane force wind on September 7. A hurricane watch was issued for Puerto Rico. But after passing Guadeloupe, Hortense encountered an area of unfavorable upper level winds known as wind shear. Maximum sustained winds dropped to 50 MPH, and the Puerto Rican hurricane watch was lifted. Most people breathed a sigh of relief, but it was premature.

On September 9 the wind shear relaxed, and Hortense began to redevelop. The National Hurricane Center computer forecast models all indicated that Hortense would continue to track west until past the longitude of our island. But since Hortense was rather close to the Puerto Rico south coast, the NHC forecasters issued hurricane warnings just in case. Most people did not take the warnings seriously.

I had serious doubts about the official forecast track however, and I made full hurricane preparations. During the day on September 9 I observed thickening high clouds moving from southeast to northwest across Rincon. A gusty northeast wind picked up but held a steady direction as the hours passed. The barometric pressure fell at an ever-increasing rate. All were weather signs known by mariners for centuries that trouble is coming.

According to the Law of Storms the wind rotates counter clockwise around Northern Hemisphere low pressure. A steadily increasing northeast to east wind accompanied by lowering, thickening clouds plus a falling barometer means that one is directly in the path of an approaching storm.

As Hortense moved along, Air Force hurricane hunter aircraft fixed the storm every 12 hours. But a late day September 9 penetration was aborted due to "extreme turbulence," likely near a developing eye wall. The NHC forecasters did not know exactly where the reforming hurricane eye was, or the storm's maximum winds. Advisories issued as late as 8 PM stated Hortense was heading west. Around 9 PM in Rincon the steady northeast wind began gusting 50 – 55 MPH, and the pressure kept plummeting. I looked to the southeast and saw continuous lightning from the approaching hurricane eye wall. Now the power began to flicker. The last words I heard from NOAA Weather Radio San Juan, Puerto Rico were "Ignore the NHC advisory. The NWS radar shows that the eye of Hurricane Hortense will cross the coast of southwestern Puerto Rico by 10 PM." By the time the forecast was corrected, the weather had become too dangerous for last minute outdoor preparations.

Screeching, keening winds reached hurricane force just after midnight September 10 in Rincon. My anemometer recorded four hours of frequently 75 – 80 MPH sustained northeast winds. The instrument could only measure up to 105 MPH. Heavier gusts often briefly pegged the needle at speeds I estimated to be 110 – 120 MPH. After 4 AM the wind abruptly slackened and shifted to the southeast at 25 – 50 MPH. The still incomplete hurricane eye wall was just offshore over the Mona Passage.

Even though Hortense was only a Category 1 hurricane, daylight revealed moderate to severe damage in my vicinity. Many homes owned by local neighbors were not well constructed, and a number of them had partial to complete roof failures. Numerous power lines were down, and I went two weeks without commercial power. I had to move my work computer to a house with surviving phone service in order to get internet access. Except for that inconvenience, I mostly made it through Hortense in good shape.

The year 1997 brought a respite from hurricanes with only one hurricane watch on a system that missed by over 300 miles. But that rest was to be a short one: then came 1998. The year began well. I saw a beautiful total eclipse of the sun on Antigua February 26. The afternoon celestial show had a backdrop of smoking Soufriere Hills volcano on the island of Montserrat 40 miles to the southwest. I felt like I was viewing the first day of creation . . . But the ancient Carib Indians as well as the Aztecs of Central America viewed eclipses as a portent of evil. I will let the reader decide.

Georges, a classic Cape Verde hurricane, formed on September 15. He quickly grew in both power and size. Sustained winds reached upwards of 150 MPH as Georges drew closer to the northeast Caribbean islands on September 19. I had very bad feelings, both professional and intuitive, about Georges as early as September 17 – 18.

My hurricane research contained plots of all hurricane tracks that ever struck Puerto Rico for more than 200 years during various months. Georges was moving right down the center line of all September direct hits. Early indications in the upper atmosphere showed nothing that would divert this extremely dangerous threat. Time for battle stations!

When facing a threat as serious as Georges, I prefer to begin preparations at least three to four days in advance of the potential blow. This is a good time to fuel up all vehicles and gas cans. Buy weeks worth of food that won't require refrigeration, store lots of drinking water, and charge up and / or replace weak batteries. Should the storm not live up to expectations, none of the early preparations go to waste. If you are subject to storm surge flooding, river or flash flooding, mud slides, or serious wind damage, preplan your evacuation routes. Contribute to the SKYWARN and ARES networks where possible. HF real time reports to WX4NHC at the National Hurricane Center are encouraged too. These can provide invaluable data to hurricane forecasters. Weather reports should include information such as station location, time, wind direction and speed, barometric pressure, and rainfall. Include reports about flooding, storm surge height, and wave height if you are in position to provide them. But remember your personal safety comes first!

If you are using a VHF repeater system, remember that repeater outages will be likely as hurricane conditions worsen. If this happens, a suggested technique is to utilize simplex operation on the repeater output frequency. This provides two advantages. One will have a better chance of finding other stations since they will be listening on this frequency. Later when the repeater comes back online, everyone will know, and they can switch over to the repeater.

Hurricane Georges maintained a relatively straight dangerous track, and I continued to elevate my preparations day by day. Fortunately a brief period of upper level wind shear weakened Georges to a still potent, large, and sprawling Category 2 hurricane as he entered the northeastern Caribbean. His first of seven landfalls was the island of Antigua.

In Rincon the daylight hours of September 21 provided the last chance to complete final preparations. That last day two friends arrived to help in last minute tasks. First item was to take down my beam yet again. Then we decided to guy wire a 450 gallon water tank on my roof. Following that we removed a 10 foot diameter C-band satellite dish and lashed it to 18 inch high rebar extending from my cement roof. There was only a small amount of glass window area to board up. Most of my house had ventanas, heavy aluminum slats that could be cranked down to keep a hurricane outside. Meanwhile the local civil defense authorities drove a bus down my street, forcibly evacuating everybody living in flimsy structures.

Just before Georges reached eastern Puerto Rico, the eye wall contracted and the clouds grew taller. The hurricane had regained Category 3 intensity. Georges took the worst possible track for Puerto Rico. In the Northern Hemisphere the strongest winds of a hurricane are generally located near the eye in the right front quadrant looking along the track that the tempest is taking. If a hurricane is moving to the west, the northwest quadrant is the worst one. Georges followed an oscillating west to west-northwest track close to the south coast of Puerto Rico. The entire island endured a prolonged period of very destructive hurricane conditions.

At sunset in Rincon my mother who lived with me asked "Where is the hurricane?" At that time the wind was northeast at 40 MPH, gusting to 60 MPH. That wind speed would be hardly noticed by someone who had spent half a century on Cape Cod. I replied "just wait." The power failed shortly, and I saw my first peak gust over 90 MPH by 8:30 PM. From that point on there were nine consecutive hours of sustained hurricane force winds, and the gusts began to peg my anemometer ever more frequently.

The very worst conditions were during the first two hours of September 22. My anemometer cups never blew away, a testament to a well-engineered instrument. Several times as the eye edged off southwestern Puerto Rico I had one minute sustained wind speeds of 100 MPH. There is very irregular terrain near my former QTH, which generated very strong wind gust accelerations. When the worst gusts arrived the noise level was equal to a jet engine whine in my back yard. I couldn't hear a human voice or the debris smashing into my home. The peak rushes of wind lasted up to 30 seconds, keeping the anemometer needle pegged. I estimated some gusts reached 150 MPH. KP4MYO located 20 miles north-northeast of me measured a 165 MPH peak gust, making my estimate credible. During the gust impacts, the solid concrete structure shook as it might during a local magnitude 4 to 5 earthquake (Puerto Rico has considerable seismic activity too). I could also observe the needle on my barometer visibly jump around in response to the extremely sudden changes in forces attacking my house. Sometimes the wind speed varied by up to 100 MPH in a matter of seconds.

Dawn on September 22: Georges moved on to the Dominican Republic and the wind rapidly fell below hurricane force. For the first time my mother and I opened the door. She said "It looks like a bomb went off." I couldn't have described conditions any better. My street was blocked by all imaginable types of debris including corrugated tin roofs, remnants of wood homes, uprooted trees, and broken power poles. Able bodied locals acted as a de facto CERT team and manually cleared away enough rubble to make the street passable. I hiked around my village to document the worst destruction I had ever seen in my life. Every wooden house along my street except one lost at least a roof. Quite a few were totally demolished.

NHC/FEMA reports later indicated that 72,605 houses in Puerto Rico were damaged significantly and 28,005 were destroyed. Georges became the most costly hurricane in Puerto Rico history. But a great deal of hurricane preparations had paid off. There were no fatalities on an island with four million residents.

You may wonder how well my storm preparations worked. I had some surprises awaiting me. My 450 gallon water tank (weight 3000 pounds when full) took flight and smashed to pieces 200 feet downhill from my place. The satellite dish, lashed to the roof, was still there. But the wire mesh in the dish was blown out by the wind, making it useless. It took 26 days to restore commercial power, and a month to get municipal water. Fortunately a gas stove, a high quality generator, and

Destruction following Hurricane Georges in 1998. [Victor Morris, AH6WX, photo]

extra water tanks at ground level saw me through the long hurricane aftermath.

The year 1999 brought some new surprises. A late October Hurricane Jose approached from low latitude waters east of Trinidad. Fortunately he narrowly missed Puerto Rico, and no hurricane force winds were reported. So was that it for the year? Based on climatology the odds were 99% that the season was over.

One factor in evaluating forecasting skill is the ability to anticipate and predict extremely rare events. In the second week of November a tropical disturbance slowly developed off Nicaragua. November Caribbean storms are not very common, and most that do develop head toward Cuba, bypassing islands farther east. So I initially had little concern over the system, which was to be named Lenny. In his first few days he moved in a slow loop over the northwest Caribbean much as expected. Then he began a course to the east-southeast then east, meanwhile becoming a full hurricane. Lenny moved well to the southeast of Jamaica on November 15.

NHC models consistently predicted a turn to the northeast or north, and hurricane warnings were issued for Puerto Rico for the seventh time in five years. By this time the constant wear and tear of making hurricane preparations had taken a toll on me. In 1999 I correctly chose to ignore Jose. But Lenny, as a unique case, was very troubling. Neither NHC forecasters nor I had any experience with a mid-November hurricane moving east across the Caribbean. I finally chose to do nothing as the steering currents near Lenny appeared to be steady from west-southwest to east-northeast near the longitude of Puerto Rico. I predicted those would bring the eye no closer than 150 miles from my home. At the time of my decision Lenny was just a Category 1 hurricane late on November 16.

Overnight tropical storm force winds gusting to 60 MPH buffeted my place. But the wind direction gradually backed from east-northeast to north, indicating that Lenny was near my expected track. But after passing well south of western Puerto Rico, Lenny finally turned more northeast and intensified explosively. Maximum sustained winds briefly reached 155 MPH in the southeast eye wall when Lenny was tangent to St Croix in the US Virgin Islands. Never before had a November hurricane strengthened so dramatically.

After passing St Croix, Lenny slowed forward speed and lost some force. He caused widespread problems in the eastern Caribbean for several days. "Backwards Lenny" became the first known hurricane to transit the entire Caribbean from west to east since the days of Columbus.

My tale ends on a whimper. My eighth hurricane warning came from Debby, a year 2000 storm. Although of Cape Verde origin, she only reached minimal hurricane status when close to Puerto Rico. The hurricane was asymmetric, and almost all the serious winds were confined to the northern semicircle. When the center of Debby was just 70 miles north of Rincon, it was just overcast and dead calm. I held an outdoor barbecue on my deck to celebrate our reprieve.

The eight hurricane warnings in six seasons for Puerto Rico proved to be a four decade record for any land area under NHC responsibility. A similar run of hurricanes occurred near the North Carolina Outer Banks 1953 – 1958.

I gained more hands-on hurricane experience in several years than many coastal residents do in a lifetime. I hope this course of events never repeats itself. But remember the lessons: expect the unexpected, be as self reliant as possible, and recognize the limitations of official forecasts. Since most of you are not meteorologists, I strongly urge you to err on the side of caution when considering hurricane preparations.

As a footnote, I moved to the northwest side of the Big Island of Hawaii in late 2001. But Mother Nature still had a few more lessons. A 25,000 acre wild fire burned as close as four houses away from me in 2005. In 2006 a close magnitude 6.7 earthquake very nearly threw me out of bed early one Sunday morning. And my tropical "friends" have not forgotten me either. We had hurricane watches from Jimena in 2003 and Flossie in 2007.

My First Hurricane, Gustav 2008

By Mike Corey, KI1U

On a Sunday afternoon in August 2008 I was contacted by the Mississippi section manager with a request for communications assistance in Louisiana. Hurricane Gustav was approaching the Louisiana Gulf Coast and was expected to make landfall on Monday morning. An emergency operations center in St Helena Parish needed communications support, preferably before the storm hit.

After confirming with the local homeland security director and getting directions I loaded up the car and set out for the five-hour drive to Greensburg, Louisiana. Arrangements had been made with another local ARES/SKYWARN member to come down three days later to assist and relieve me. Despite being active as a storm spotter for many years and having seen tornadoes, floods, blizzards, and ice storms I had never experienced a hurricane before, so nerves were running a bit high.

Having been involved in ARES and SKYWARN for sometime I had a basic "go-kit" ready to go. I added a few other items — an HF rig, base 2 meter antenna, and a G5RV antenna — and thought I was all set. Shortly after the storm made landfall I had a running list of "should have brought" items.

Upon arrival I was briefed by the local homeland security director and began setting up radio equipment. It was almost midnight when I arrived and too dark to set up antennas. The forecast called for Gustav to make landfall at 9 AM Monday morning, so we thought this would give us enough time to finish the antenna installation at first light. At 7 AM there was a briefing via WebEOC with the Governor's Office of Homeland Security and Emergency Preparedness (GOHSEP). Following that last minute preparations were made. Due to a shortage of available hands a temporary HF antennas was deployed. A Ham-Stick type antenna was mounted on a cookie-sheet with some wire radials and placed on the roof of the EOC. It didn't work great but allowed us to stay in touch with GOHSEP. At the last possible minute the G5RV was put up.

Assisted by local public works employees the antenna was placed on a 35 foot tower. At the time of its installation the rain had started and the wind had kicked up to 60 – 70 MPH with gusts considerably higher. Despite being drenched we were on the air. At approximately 9 AM Gustav struck and claimed its first victim, the Ham-Stick. The winds had bent it in half and blown it off of the roof!

Not long after this the winds damaged the doors to the EOC. Until the storm let up it was a full time job keeping the water out. An hour after landfall the EOC generator failed. It wasn't until three days later that we were able to get a repairman out to fix it. To get us through until the generator could be brought back online two small portable generators were brought in. One ran the lights on the EOC floor and the other provided power to the radios. No lights and lots of water in the building presented just a few of the problems we had to contend with. It was also during the first few hours of the storm that cellular communications failed; first voice, then text.

Immediately following landfall, rain was the main thing we dealt with as far as weather was concerned. In all Gustav dumped around five to seven inches in that area, while other nearby parishes received more than 15 inches. Wind was another factor, causing trees and power lines to fall blocking roads. As Gustav made its way inland reports of tornadoes started to come in, the first was received around 10 AM.

The first day we received nine reports of tornadoes in the parish. As the reports came in to the communications room they were relayed to the EOC floor and to the National Weather Service. On occasion relaying a report to the nearest NWS office took a circuitous route. Around 9 PM on Monday evening a report came in of a tornado in the southwest corner of the parish. Attempts were made to contact NWS Slidell and Jackson, but were not successful. Contact was finally made with the amateur radio station at NWS Dallas-Ft Worth. Within a matter of minutes a tornado warning was issued for St Helena parish. Despite the roundabout means of communication we were able to stay in touch with the NWS and receive periodic four-hour forecasts.

By the end of the first day there was a lull in activity that allowed for EOC staff to rest. This little down time even allowed for a quick QSO with UA0SJ in Asiatic Russia on 20 meter CW (storm spotters are DXers too!) Not long after midnight, though, the rest came to an abrupt halt when a report came in of a gas leak. A request was made to get a message through in a hurry to the gas company to get a crew out. The situation was critical because the area with the gas leak was about to go underwater. A message was passed via the 3872 tactical net to the gas company. The crew eventually arrived and took care of the situation. Around 2 AM everything was stabilized so EOC staff could get some sleep.

The second day began with more reports of tornadoes in the area. Tornadoes caused by hurricanes are a bit different than the tornadoes usually associated with thunderstorms. Hurricane-induced tornadoes, which generally appear in the upper right quadrant of the storm, are usually short lived and weak (rarely above F1). Ground truth reports may be hard to come by for several reasons: spotters may not be able to safely get out and observe, the

The back side of the emergency operation center in Greensburg, Louisiana, showing the tower and G5RV. [Mike Corey, KI1U, photo]

Rapid Comms trailer that provided satellite-based Internet and phone service. [Mike Corey, KI1U, photo]

tornado may be rain wrapped, and it may not last very long. Tornado warnings after a hurricane makes land fall may not be accurate because of these factors, but it is better to err on the side of caution.

The second day also marked the beginning of a recovery phase for the area. There were needs in the parish for fuel, food, water, and ice. The EOC needed the generator fixed, more 700 MHz radios, phone and internet, and security personnel. The staff of the EOC was fairly small so many had to wear more than one hat. My role transitioned from a radio operator to assisting the local homeland security director as his deputy. Throughout day, two requests were submitted for the needed commodities. Also it was a time for EOC staff to send messages out to family and friends to let them know that they were OK. The first to arrive was the Louisiana Air National Guard with a Rapid Comms trailer that provided satellite internet and phone service. Despite numerous requests, the security personnel (Army National Guardsmen) had not arrived by the end of the second day. We needed to have them arrive before the fuel, food and ice, so time was of the essence.

Day three we found the proverbial needle in a haystack, a generator repairman. After several messages were passed through the tactical net one was found about an hour away in Baton Rouge. By midday the generator was back up and running. This made an internet connection possible again, so the Air National Guard unit with the Rapid Comms trailer could pull out to be deployed elsewhere. There was still no sign of our security team or the supplies we had requested so follow up requests had to be made. And as shelters filled up amateur radio was put to use relaying information for the Red Cross. Communications assistance was also given to local health department officials and social services. Day three was also the arrival of the first relief amateur radio operator, Doug, K5DSG.

An interesting challenge for day three came from the media. Local media were reporting that supplies were available from the EOC and people just needed to stop by. There were two problems with this. First, we hadn't received any supplies by this point. Second, if we had any supplies we would not distribute them from the EOC for security reasons. During Hurricane Katrina a similar problem occurred. A local 911 center determined that in such a situation the communications center is better suited to set the local media straight instead of going through normal public relations channels that might be tied up. We employed that lesson learned and stayed in contact with local media so they could give accurate reports. Since the EOC staff was small this freed up the director to handle other issues.

By the end of the third day our requests for supplies were being filled. Starting around 11 PM supplies started arriving: ice, oxygen tanks, 700 MHz hand held radios, security personnel from the Tennessee National Guard, personnel to help distribute supplies from the Kentucky National Guard, bottled water, MREs (Meals Ready to Eat), and fuel. With supplies now in hand a plan for distribution could be made.

Day four was to be our final day in St Helena, Louisiana. David, KE5STF, and Peter, KE5STL, arrived to assist at the Red Cross evacuation shelter. Distribution of food, ice, water and fuel also began. The plan for distribution worked extremely well. People lined up along a northwest to southeast road north of the distribution point. They then turned onto a road that led south to the distribution point and picked up supplies. Each person was allowed to receive a set amount of food, ice, water and fuel. They then proceeded south to another road where they could turn east and head back into town or west and head into the parish.

After passing some final messages for the Red Cross, Department of Health, and Social Services we verified that the communications situation was stable and nothing else needed. By 2 PM we determined "mission complete," packed up, said our farewells, and hit the road for home.

Many lessons were learned during those five days in St. Helena, Louisiana. In retrospect I am certain that we were not the first to learn these lessons. Here is what we learned from our experience.

Go in with a mission plan. The chance to assist when needed is quite exciting and in many places having extra personnel is very welcome. But a volunteer who shows up to help without a plan or set of goals can quickly become a catch-all for every imaginable task. Before you go, establish what your goals are, a time-frame to get them accomplished, and know your limitations. In our case we had a three-point mission plan: set up amateur radio communications, provide communications while assisting in re-establishing normal communications, and make sure communications are stabilized before departing. Admittedly we undertook more than what our plan outlined: media relations, EOC management, and supply requisition. Through my experience working in an EOC I was comfortable handling these tasks and at no point felt that I was outside of my ability to do the job. Not everyone would be able to do this. It is still best to stick with the plan and not vary too much from it if at all possible.

If you find yourself in a disaster such as a hurricane plan for being self sufficient for five days. Despite being in a well-equipped EOC we were without power, cell phone communication, internet, shower facilities, and had only the supplies on hand for the first few days. Having a well supplied go-kit will help greatly in getting by those first few days. Also include in your go-kit a small portable weather measuring device that can tell temperature, barometric pressure, and wind speed. This will allow you to monitor weather conditions and report them, if needed, to the NWS.

There is a definite pattern to cell phone outages. Voice communication is the first thing to go since it requires the most bandwidth. Text messaging is about the last thing to go since it requires very little bandwidth, so if you can't make a call try sending a quick text message. The message may not go through immediately but most phones will make repeated attempts to send the message.

Training is an absolute essential for responding to disasters. At the basic level you should have completed SKYWARN or a comparable storm spotter training course and the ARRL Emergency Communications Course. Beyond that online courses available through FEMA are of great value to disaster responders, especially EOC, working with volunteers, and NIMS courses.

In responding to St Helena we found ourselves working, in a disaster, alongside people we had never met and knew

Amateur Radio operating position showing radios for HF/VHF/UHF Amateur Radio communication, satellite telephone, and a 700 MHz radio that linked emergency operations centers in the area and the Governor's Office of Homeland Security and Emergency Preparedness. [Mike Corey, KI1U, photo]

nothing about, in a town we had never been to. And from their perspective we were two strangers and they knew little to nothing about us. There is a story that is worth sharing. One of the great college rivalries is between the University of Mississippi (Ole Miss) and Louisiana State University (LSU). I was employed and attended Ole Miss and showed up in St Helena Louisiana donning my Ole Miss Baseball hat. The EOC was staffed by diehard LSU fans, and many grumbled about being sent an Ole Miss fan to assist them. Over the course of our time there a good working relationship was built and I showed that despite being an Ole Miss fan I could be trusted. Several months later while passing through the area, coming back from New Orleans, I stopped by and visited with the local homeland security director. After comparing notes on our experience during Gustav I presented him with his own Ole Miss Baseball hat. He said that is one souvenir that would get a prominent place on his desk.

During a hurricane many volunteers come from outside the affected area and find themselves in the same situation. It is absolutely critical that you have good people skills and not lose sight that you are not only a volunteer but a guest in their community. It is important that you show that you can be trusted and can work as part of the team. Keep in mind this team may have been working together for many years so this sounds easier than it really may be.

From a storm spotter's perspective a hurricane offers several serious challenges. During normal spotter activation there are mobile spotters that report conditions that they observe. During a hurricane mobile spotters most likely will not be available, making ground truth reports hard to come by. Many reports that we received came from the general public and most likely were from people not trained in observing severe weather. Contact with the nearest NWS office may also be a challenge. Lines of communication may be routed through other places, such as our case contacting the Dallas-Fort Worth NWS office. Getting an accurate forecast is also more difficult. We would get four-hour forecasts but they were far less detailed than we would have preferred. And forecasting exactly where a hurricane will make landfall is just not possible. Two hours before Gustav made landfall it could still not be pinpointed on a map.

Response to a hurricane is definitely a combination of storm spotter response and emergency communication response. If you respond and neglect the storm spotter aspect you could find yourself underprepared.

Hurricane Katrina

Ham Radio Operations in Mississippi

By Malcolm Keown, W5XX

Mississippi is in a unique climatological location resulting in a variety of weather events. In the winter frigid air from the north collides with warm Gulf flow which can result in severe ice storms. Spring brings warmer weather but the same type of weather systems with sharp temperature differentials resulting in tornadoes and widespread damaging thunderstorms. As the summer ends the southern coast of Mississippi is subject to hurricanes while the northern half of the state can be traversed by tropical force winds and rain. The result is that Mississippi radio amateurs need to be ready for emergency response the year around.

In recent times the Mississippi coast has been impacted by two catastrophic storm events — Camille in 1969 and Katrina in 2005. Most hurricanes form off the western coast of Africa as tropical waves and later organize into low pressure systems as they move across the Atlantic. Katrina was different. This storm formed over the Bahamas on August 23, 2005, and crossed southern Florida as a Category 1 hurricane, resulting in deaths and flooding before reconstituting over the Gulf of Mexico into a Category 5 hurricane. As the storm approached the Gulf Coast it weakened to Category 3, making landfall near the mouth of the Mississippi River, and then moved north across the marshlands east of New Orleans, making its final landfall in Hancock County, Mississippi.

Due to the counterclockwise rotation of the storm, the Mississippi coastal counties were pounded by 120 MPH winds and a devastating 25 to 30 foot storm surge which penetrated as much as 12 miles inland. Just to the west of Hancock County, the city of New Orleans was inundated from the north as the winds rotating around the low pressure area to the east of the city raised a storm surge across Lake Pontchartrain that pushed devastating floodwaters into New Orleans. After landfall, Katrina traversed the length of Mississippi as a Category 1 hurricane and later as a tropical storm exiting the state on its northeast corner some 400 miles to the north.

Although the ham population in Mississippi is thin, considerable preparation had been put in to motion by ARES members to deal with storm events. These hams had practiced operating under emergency conditions during Field Day, coordinated with served agencies through planning for the annual Simulated Emergency Test, regularly checked into nets to understand net protocol and how to pass formal traffic, and taken the ARRL Emergency Communications Courses to understand emergency procedures and organizational structure. In addition, Mississippi had signed assistance Memoranda of Understanding with surrounding ARRL Sections to operate a net to handle tactical traffic, leaving the traditional "Mississippi frequency" of 3862 kHz open for the Magnolia Section Net and Mississippi Section Phone Net to handle health and welfare (H/W) traffic. This was fortuitous during Katrina in that many of the experienced Mississippi net control stations (NCS) were in the southern part of the state and were off the air. Fortunately, NCS primarily from Arkansas, Louisiana, and Texas stepped up to help keep the tactical and H/W nets in operation.

By the time Katrina had passed out of the state hams were short on sleep, had been wearing the same clothes too long, had too many flashlight batteries that were shot, and had gas gauges and cans that were empty. Realizing that there was absolutely no communications on the coast, hams rose to the occasion and activated W5SGL at the Harrison County EOC and the KC5IF 146.73 repeater (on battery power) which was the only repeater still intact. Gulf Coast District Emergency Coordinator Tom Hammack, W4WLF, noted that for the first three days after Katrina made landfall, amateur radio was the only functional communications on the Gulf Coast.

As a result, W5SGL through KC5IF/R coordinated emergency communications across the Gulf Coast for local fire, rescue, police, and ambulance services, and arranged logistics for medical evacuations, feeding, heavy equipment movement, and military operations. In addition, operators at W5SGL expedited press releases and health notices, as well as handling traffic for the Coast Guard, FEMA, the Mississippi Emergency Management Agency (MEMA), and missing persons and disaster mortuary teams. The W5SGL log shows that over 100 trapped victim rescues as well as numerous ambulance dispatches for medical emergencies resulting from operations at the Harrison County EOC ham station.

As the situation began to stabilize other hams managed to get on the air and help with many types of tactical traffic as well as with H/W traffic. Mississippi hams to the north of the Gulf Coast managed to link the Gulf Coast by 2 meter repeater to MEMA in Jackson, providing a vital link for passing traffic. In addition to providing emergency communications, hams also assisted in the disaster relief effort in other ways including helping in shelters, participating in search and rescue operations, clearing roads and utility rights of way, and securing and delivering needed supplies. Hams were truly willing to go the extra mile in spite of family needs, enormous personal losses, and professional and job commitments.

Humanitarian assistance reports due to the efforts of hams were numerous. One particularly interesting report came from Forrest County Emergency Coordinator Lex Mason, KD5XG, of Hattiesburg, who worked at the Forrest County EOC during Katrina. The request that really stuck in his mind was a call from Camp Shelby, a National Guard Training Base, just south of Hattiesburg. A soldier in Iraq called Camp Shelby officials and requested help in contacting his father, who lived by himself and, if not found, would die for lack of medical care. Lex was very concerned about the soldier's dad and the remote area where he lived. The directions included going to Lucedale, Mississippi, turning south on Highway 57, crossing the Jackson County Line, and then turning west. Additional instructions were given about passing a cemetery and church. He lived on a dead-end road somewhere in the woods past the church, all very confusing to say the least. Through the directions received, Lex surmised that the man was probably in Jackson or Harrison County. The devastation in these two counties was unbelievable. There were no communications in the area and many roads were impassable due to downed tress. Not knowing which county the man lived in added to the difficulty.

Total devastation in Waveland, Mississippi. [National Weather Service]

Fortunately, the soldier requesting the assistance later gave a Perkinston, Mississippi address for his dad, which would put him possibly in Stone County. KD5XG was able to make contact with Sarah Purvis, KC5NFI, in Stone County. She delivered the message to the sheriff's office. Within two hours Sarah reported that the dad had been located and was safe. The contact at Camp Shelby was advised that the soldier's father had been contacted and was OK. Lex said that helping this one soldier locate his dad and giving him assurance that he would receive the necessary medical care was worth all the work devoted to the Katrina emergency operation. he further said that our military service personnel put their lives on the line for us daily and helping one of them was indeed a privilege.

After Katrina had passed and Mississippi hams had a chance to get their lives back into order, a number of them met in Hattiesburg to discuss the lessons learned from this catastrophic event. These lessons were summarized as follows:

(a) Government and non-government agencies should not attempt to activate any ham radio emergency communications response without proper ARRL

Views of inundated areas in New Orleans following breaking of the levees surrounding the city as a result of Hurricane Katrina. [National Weather Service]

Section-level ARES coordination. This is not to say that emergency communications assistance was not needed during Katrina; however, there was no ham radio national/regional emergency response plan in place, thus the uncoordinated inundation of the Gulf Coast and South Mississippi with assisting operators was unexpected and unprecedented. In that current ARES policy is to organize emergency communications response on a county/district/section basis, the uncoordinated action resulted in cases of confusion, inefficiency, and hard feelings. Some outside operators seemed to think they were in charge of local ham radio personnel and resources (repeaters) regardless of the fact that an ARES emergency response structure was already in operation.

(b) A pre-populated database of emergency communications volunteers and rapid response teams (RRTs) should be maintained. The well meaning on-line method used during Katrina resulted in enlisting many helpful operators; however, many operators showed up for assignments who were ill-trained, unprepared for a stressful assignment, and had personality problems not compatible with the disaster environment. We don't need "fire engine chasers" and "cowboys." This database should be actively updated. This database should be populated by nationwide ARES Section-level leadership recommendations, and these recommended responders must be properly credentialed.

(c) Emergency Coordinators should be in charge of ARES operations in a given county/jurisdiction provided the EC is properly coordinated with the county emergency management agency (EMA) director. In most Mississippi counties the EMA Director works under the county Board of Supervisors. The EMA Director appoints a communications officer, who may or may not be the EC. In any case, the EMA Director should be aware of all emergency nets in operation in his jurisdiction including an ARES net. In the event the EC is also the Communications Officer, he should be aware of all net operations in a given jurisdiction and be prepared to report the current status of these operations to the EMA Director at any time. There is no problem with having separate ARES, Red Cross, Salvation Army and Baptist Men's Kitchen nets in operation: the key point is that they should all properly be coordinated with the Communications Officer so that communications resources are efficiently utilized.

During the course of the Katrina landfall and following response, an article in the Wall Street Journal referred to ham radio as a "hobby" similar to stamp collecting. Clearly the emergency communications support provided by amateur radio operators during Katrina demonstrated that we are truly a "service" and not just a "hobby."

Hurricane Sandy
By Vic Morris, AH6WX

The Mid Atlantic to New England area is mostly not known for a high hurricane risk, although destructive tropical cyclones are rare visitors. Typically the highest risk period in this region is the last half of August and September. The majority of hurricanes in the western Atlantic turn northeast and pass seaward of the Carolinas and points northward. Once in a while one directly impacts the Carolinas or New England. Storms in these latitudes nearly always accelerate on a course somewhat east of north. The cause is the typical mid latitude westerlies found in the middle and upper troposphere. Hurricanes crossing the Carolinas moving north to northeast quickly lose strength if they spend any significant amount of time over land. The typical greatest danger from a tropical cyclone in the Mid Atlantic area is fresh water flooding from excessive rainfall.

However Hurricane Sandy, a major rule breaker, brought death and destruction on a wide swath from the Caribbean to Canada in late October 2012 with particular emphasis on New York and New Jersey. As of 2014 total damage was estimated at $68 billion. This makes Sandy the second costliest hurricane only behind Katrina of 2005. There were 72 deaths in the United States, the deadliest US cyclone north of the Carolinas since Agnes of 1972. Agnes produced tremendous fresh water flooding near Pennsylvania.

Sandy started in routine fashion, as a tropical wave moving off the coast of Africa in the second week of October. This is near the end of the time when significant tropical development is possible in the Cape Verde area off West Africa. The tropical wave moved westward but did not develop for a long time due to sinking dry air aloft. However by the time the tropical wave reached the central Caribbean on October 20, it encountered moist, unstable, rising air and the pressures began to fall. A defined tropical depression formed on October 22 south-southwest of Jamaica. It turned north-northeast in response to an upper trough over the Gulf of Mexico and northwestern Caribbean. The system became Tropical Storm Sandy six hours after reaching depression status. Sandy became a hurricane the next day, crossing eastern Jamaica as an 85 MPH Category 1 system. From there Sandy rapidly intensified, entering Cuba pre-dawn October 25 near Santiago de Cuba with 115 MPH (Category 3) peak sustained wind. Up through this time the tropical cyclone had a typical fairly small diameter of damaging winds.

Sandy took five hours crossing Cuba, and emerged into the southwest Atlantic as a Category 1 hurricane. From this point, things became complicated. A sharp trough with a negative tilt (oriented northwest to southeast) was located over the southeast USA and adjacent ocean. This bent Sandy back to a northwest track. Sandy moved along the outer Bahamian Islands roughly from Eleuthera to Great Abaco. Some cooler drier air encountered the fringes of Sandy forming a frontal structure. This is a sign that Sandy was starting the process of becoming an extratropical storm, one fueled by thermal frontal contrasts. Briefly the core of Sandy weakened to a tropical storm after passing the Bahamas on October 26. At the same time the circulation more than doubled in size. Wind gusts over 60 MPH were noted on buoys off southeast Florida.

The trough steering Sandy northwest weakened some, and Sandy resumed a northeast track over the southwestern Atlantic, regaining hurricane force. The extratropical transition was not completed and thunderstorms fired back up near the hurricane core. At this point it is interesting to compare the major computer model forecasts on Sandy. As early as Sandy became a named storm, the ECMWF (European model) predicted a landfall in the northeast USA, although many details on intensity and size were uncertain. However the GFS (American model) initially expected Sandy to go out to sea and not come close to land after leaving the Bahamas. But over time the GFS model aligned with ECMWF solutions calling for a Northeast or Mid-Atlantic states landfall.

Next a piece of an upper trough located in the central USA moved southeast. Meanwhile a most unusual upper level high was located over Newfoundland. This set up a southeast to east steering current over much of the northwestern Atlantic, and Sandy began to move northwest about 20 knots towards shore. Meanwhile wind shear above Sandy's core was light, and the hurricane traversed abnormally warm waters of the Gulf Stream. Maximum sustained winds peaked again at 100 MPH. Also extratropical frontal processes renewed along the edges of Sandy. The area of tropical storm force wind expanded to 1000 miles, a record for an Atlantic hurricane. Finally as Sandy neared the New Jersey coast late October 29, cold air reached the storm core and deep convection near the center ceased. Thus Sandy was declared post tropical just prior to landfall, but peak sustained winds were still 80 MPH. At the time of crossing the coast, a tide gauge station near Atlantic City reported a minimum sea level pressure of 945.5 millibars, an all-time low near this region. After moving inland Sandy moved west-northwest and slowed down while gradually weakening. The remnants finally drifted into Ontario, Canada by November 1.

Sandy impacted a huge area from the Caribbean to the Canadian Maritime provinces and westward to the Great Lakes region. The primary impacts along the East Coast were damaging winds and storm surge. The highest measured wind speed was 75 MPH at Great Gull Island, New York on Long Island. The peak gust reported was 97 MPH. Winds gusted to hurricane force in seven states from New England to the Mid-Atlantic States. A very respectable 68 MPH gust was observed at the Lakefront Airport in Cleveland, and gale force gusts reached as far away as Wisconsin.

Since most of the coastal damage was surge related, it is worthwhile to define a few terms. All coastal locations have astronomical tides produced by the moon and the sun. The tidal ranges are greatest near new moon and full moon. A storm surge is the rise in sea level produced by the combination of persistent onshore high winds over a long expanse of water, low pressures, and bathymetry (the shape and slope of offshore bottom contours.) The storm tide is the sum of the astronomical tide plus the storm surge. Meanwhile actual inundation of normally dry ground is measured by the difference between the height of the storm tide and the mean high tide elevation. The inundation is easiest for most persons to visualize as that water height is what impacts low lying coastal structures.

Sandy's peak surge had the misfortune of arriving on a peak full moon fall high astronomical tide at many locations. The winds blew directly down Long Island Sound, bottling up sea water at the west end of the Sound. A tide gauge at Kings Point reported a 14 foot total storm tide. The Battery near New York Harbor recorded almost 10 feet. In terms of inundation the worst impacted locations of Staten Island, Manhattan, and northern New Jersey reached up to 9 feet deep. Some exposed locations from southern New England to northeastern North Carolina had several foot inundations. The water rises near New York City and New Jersey set many all-time records. The exceptionally large area of onshore tropical storm to hurricane force wind created surge heights typically found only in more intense hurricanes.

In New York the surge damaged or destroyed 305,000 homes. Damage in New York City alone was estimated to be $19 billion. Portions of the subway system were flooded, and there was no service between Brooklyn and Manhattan for several weeks. In portions of New Jersey whole communities were inundated, homes washed off foundations, boardwalks buckled, cars tossed, and boats were thrown well inland. Natural gas lines ruptured in places setting fires. 346,000 homes were damaged, and 22,000 of them were declared to be uninhabitable. At the height of the storm 5 million customers lost power, and it was common for outages to last several weeks.

Sandy had other impacts. It was a heavy rain producer, although not much fresh water flooding occurred. A peak USA total of 12.83 inches fell in Bellevue, Maryland. Cold air entered the western part of post tropical Sandy and blizzard conditions prevailed at higher elevations of the Appalachian Mountains. Snowfall totals over a foot were common especially in West Virginia. Two stations reported 36 inches.

Evacuations and preparations for Sandy could have been better. Most of the forecasts for Sandy were timely and accurate. However the National Hurricane Center had no procedure in place for coastal warnings on a hurricane turning post tropical. Tropical storm warnings were issued only up to North Carolina. Local National Weather Service offices farther north issued high wind warnings for their areas of responsibility inland, and hurricane wind or storm warnings for offshore waters. It was felt by some that the local warnings did not convey the same sense of urgency as a National Hurricane Center hurricane warning. In a post storm review the National Weather Service changed their policy. In future similar events the National Hurricane Center will issue appropriate watches and warnings for any post tropical cyclone that remains a threat to life and property. This will maintain the continuity of warning information during the transition from a tropical cyclone to a post tropical cyclone.

Hurricane Sandy 2012
By Jim Mezey, W2KFV, ARRL New York-Long Island Section Manager

The ARRL New York City/Long Island Section is a very diverse section. There are three Districts, each with their own District Emergency Coordinator. One DEC covers New York City, which is mainly urban, consisting of five boroughs: Queens, Kings (Brooklyn), Manhattan, Richmond (Staten Island) and the Bronx. Population is approximately 8.5 million. One DEC in Nassau County which is suburban and urban, has three townships with a population of approximately 1.4 million. One DEC in Suffolk County, which is mostly rural, has 10 townships and approximately 1.5 million people. There are Emergency Coordinators for each borough in New York and each township on Long Island. As you can see this creates unique challenges for the NLI Section. Each District Emergency Coordinator is familiar with their respective area and its resources and needs.

Most activations of amateur radio in the NLI Section are due to weather related activities such as winter nor'easters, blizzards, tropical storms, and hurricanes. The NLI Section is 135 miles long including Staten Island, and 23 miles wide at its widest point. The whole Section is surrounded by water with Long Island Sound to the north, Atlantic Ocean to the south and east and the Hudson River, Arthur Kill and Kill Van Kull (both tidal straits) to the west.

Hurricane Sandy proved to be a powerfully impressive storm in terms of its danger and strength, ranking as the second costliest named storm in US history. It ranks in severity only second behind Hurricane Katrina (2005). Hurricane Sandy originally formed on October 22, 2012 as a tropical depression in the southern Caribbean Sea off the coast of Nicaragua. The depression then strengthened to become Tropical Storm Sandy, with maximum winds of about 40 MPH. On October 24 Sandy became a Category 1 hurricane as it moved northward across the Caribbean and crosses Jamaica with

Damage from Hurricane Sandy in Breezy Point, Brooklyn [Russ Logar, KC2LSB, photo]

winds of 80 MPH. Although Sandy's eye did not cross the Dominican Republic and Haiti to its east, the storm dumped more than 20 inches of rain on Hispaniola. More than 50 people died in flooding and mudslides in Haiti. October 26 Sandy strengthened as it moved from Jamaica to Cuba and struck Santiago de Cuba with winds of about 110 MPH, only 1 MPH below the status of a major Category 3 hurricane.

As Hurricane Sandy started heading toward the New York City area, all ARES personnel were advised to make preparations according to the plans they have trained with. All equipment was checked and rechecked. There were many meetings with local government personnel as well as served agencies. Red Cross requested amateur radio support for 25 possible shelters throughout the Section, with the largest shelter having a 6000 person population. The fire service in Nassau County, a served agency, consists of 71 departments divided into nine battalions, asked for ARES personnel to assist at the nine battalion emergency operation centers. Nassau County Office of Emergency Management also requested amateur radio support.

Most townships throughout Suffolk were asking about amateur radio support availabilities for their EOCs. Forecasts called for Sandy to downgrade to a post-tropical cyclone and make landfall in New Jersey. Despite the downgrade in classification a catastrophic impact was still expected. On Saturday, October 27, two days before the storm made landfall, the National Hurricane Center downgraded Sandy forecast to a post-tropical cyclone. The Center decided that it was best to leave the warnings to local officials. Sandy made landfall just north of Atlantic City, New Jersey on October 29 at 7:30 PM and drove a catastrophic storm surge into New York (and New Jersey) coastlines.

SKYWARN nets were active giving updates of the storm's progress up the coast and into the region. The Resource/Information Net was also activated. This net is run by the Long Island Mobile amateur radio Club. They used a linked repeater system that covers most of the New York City/Long Island Section. This net was designed a few years earlier during the Northeast Blackout, to gather and pass information by mobile ham commuters of road hazards, closures, open gas stations, flooded roads, downed wires, trees, auto accidents and so on. They also gave storm updates and progress reports of phone and power service. This proved to be invaluable during Sandy and in the initial recovery period.

As Sandy progressed, conditions in the three Districts degraded dramatically. We were lucky that the wind fields were broken up and not as bad as anticipated; however, we still had a large storm surge that created a lot of inland flooding, especially in the Back Bay areas of the south shore areas of the Section. RACES was operational in Suffolk County, while RACES/CERT Communications in Nassau County were operational at the Office of Emergency Management.

We lost more than half of our ARES operators due to evacuations, downed trees and wires, flooding and fires on the south shore of Brooklyn and Long Island. There were over 1.7 million without power on Long Island alone, including many amateurs. Most phone service was nonexistent and cell phones were overloaded and failed. This required a quick rethink in operations. Information received from emergency management and the Red Cross indicated the population of shelters was not as great as originally thought and that most of the general public insisted on sheltering in place. We made some adjustments. District Emergency Coordinators and Emergency Coordinators reevaluated what they needed and adapted to serve their communities as well as town and local governments. Additional District "Update Nets" were implemented to augment anticipated needs throughout the Section. Many clubs as well as ARES groups helped their local communities with transporting needed goods such as food/water and donated items. They also set up charging stations for cell phones and mobile devices. We prepared to look outside the NLI Section for amateur radio mutual assistance if necessary. We were very lucky that many repeaters that we use were not affected by this storm.

During the days after the storm's passing, damage assessment was conducted and communication needs were reassessed and assets deployed as required. Many radio amateurs worked long, stressful shifts and were relieved to get some much needed rest.

We continued with the Resource/Information Net for the next week or so, continuing to provide updates on areas with or without power, stations to find fuel and so on. Fueling vehicles had become very problematic as most stations had either run out of fuel or had no electricity to power their stations. The fuel terminals in the Island Park area of Long Island had no power and quickly depleted their reserves of bulk gasoline. The infrastructure on Long Island had collapsed with raw sewage backing up into the streets as well as being dumped into the back bays of the south shore. This had created a major health risk to all first responders as well as the homeowners. Debris fields were large and travel was restricted. Power lines were down for weeks.

The National Traffic System was very busy moving health and welfare messages. The Red Cross set up "Food and Clothing Reception Areas" throughout the Section. These were staffed with ARES radio personnel until regular communications became available.

On October 31 Sandy dissipated over western Pennsylvania, and the National Weather Service issued its final advisory on the storm. NWS's advisory said "multiple centers of circulation in association with the remnants of Sandy can be found across the lower Great Lakes."

When we thought we were done and we could start getting things back into normal operations Mother Nature decided she was not done. To add insult to injury, one week later a blizzard dumped more than 8 inches of snow across the city and the island.

Amateur radio ramped up again with SKYWARN and Resource/Information Nets transmitting information which was beneficial to our first responders and served agencies. In the days that followed ARES and other amateur radio groups stood down. All the served agencies thanked the amateur radio community for their cooperation and assistance for the safety and welfare of the public.

Some of the lessons learned from Hurricane Sandy are useful to amateur radio groups.

• Establish a Resource/Information Net separate from operations or administration nets, to get on-the-spot information that can be passed on to those who may be able to utilize it.

• Twelve hour shifts seemed to work the best. While this may seem like a long shift it was similar to shifts worked by other volunteers and was the best possible with the manpower available. Eight hour shifts would be more feasible but were not practical.

• Operators living in metro areas may have difficulty getting around due to mass transit shut downs. Operating from home is an alternative and should be done so in coordination with support activities at the served agency.

• Consider an ARES reserve group. This would be a pool of operators who want to help out during storms and other activations but are not available as full time ARES members for various reasons. In Nassau County, these hams are given a Field Operating Guide and have regular online and in-person training to keep their skills sharp.

• Get radio clubs involved to help out in their communities. Talk to them before the event. Establish and set up cell phone charging stations, internet connections and so on.

• The ARESMAT concept for resources works well. As a Section Emergency Coordinator or Section Manager you may have to plan for housing if you are bringing outside amateur radio resources for long periods. What about food and water? Who is going to fund it? You can contact ARRL Headquarters for assistance with this and other needed resources.

• Do not over extend yourself with served agencies. Realize you may not be able to get many of your responders to the event, especially if it is a weather event. Have a contingency plan.

• Consider having written Emergency operation plans for events that happen in your areas. Update them as necessary.

• Recheck phone numbers and email addresses at least once a year to make sure they are still operational.

• Train for proficiency, use your "Go Bags" and keep your batteries charged. Familiarize yourself with the resource of Ham-Aid (available from ARRL HQ through your section manager). If

This image from the NASA/NOAA Suomi NPP satellite's VIIRS instrument shows Hurricane Sandy strengthening in the Gulf Stream as it makes its way to landfall in New Jersey. The image shown here was taken during the satellite pass around 1735Z on October 29, 2012. [Courtesy NOAA]

additional equipment is needed for the response, get the request in before the storm, not during or after.

Hurricane Maria 2017

After-Action Report from the ARRL Puerto Rico Section

By Oscar Rest, KP4RF, ARRL Puerto Rico Section Manager

Radio amateurs were very active during the emergency of Hurricane Maria from September 20, 2017, until the end of November as normal communications were reestablished. This Category 5 hurricane devastated the island, affecting the normal life of 3.4 million US citizens. From the catastrophic impact of Hurricane Maria, we suffered a power outage of 100%, with a communication loss of 98%.

We have documents, situation reports, and press publications that record the active participation of the amateurs radio operators. These reports contain messages of requisitions for hospitals, mayors, Power Authority (AEE), and emergency management among others. These requests were channeled through the Disaster Relief Operation (DRO) of the American Red Cross (ARC) in Puerto Rico and directly through FEMA Join Force Operations (JFO).

Among the messages transmitted/relayed for emergency requisition were medicines, diesel fuel, power generators for hospitals and the AEE Technical Plazas, and even drinking water. There was also the possible collapse of the Guajataca dam that could flood valleys, putting in danger thousands of citizens. Also, we communicated the coordination of the transfer of intensive care unit patients between hospitals, and the reunification between Puerto Rico citizens and their families in the continental United States, where the ham radio amateurs covered the communications gaps in the disaster region of Puerto Rico.

During the catastrophe, the Power Authority (AEE) lost all communications. We developed a compromise with the Power Authority to provide emergency communication from Monacillo Control and the Technical Plazas. The amateur radio operators provided tactical and technical communication for the reestablishment of the electrical infrastructure on 23 Technical Plazas located at electrical grid nodes scattered across the Puerto Rico territory.

By September 23, after the hurricane ended, radio amateurs had "eyes and ears" on the whole island on VHF frequencies, where we created an incident map following the Memorandum of Understanding (MOU) signed in April 2017 between the ARRL Puerto Rico Section and the American Red Cross local chapter. This incident map contained the assessment of the road conditions and the descriptions of the Hurricane Maria's high-level damage over Puerto Rico by the communications from the AEE Technical Plaza operators and the drivers of Power Authority service trucks.

We copied reports of the need for water, diesel, power poles, wires, transformers, gasoline, and food for the operators. This situation appraisal allowed us to analyze how to best deploy our amateur radio communication resources. We knew how to get from one town to another as we received reports of roads and traffic obstructions.

The American Red Cross was able to follow this incident map to deploy their teams, and they also presented the map to FEMA. Immediately the Emergency Support Function (ESF-2) requested that the local ARRL Section Manager (SM) report and participate in their daily meetings and help to reestablish the communications across the island as we were the only communication available in the hurricane-impacted region. In addition, we were requested to cover the communication needs of the 11 main hospitals across the island.

The active participants were 131 local radio amateurs that operated at the AEE Technical Plazas, and shortly after Maria more than 24 specialized amateur radio operators arrived under the American Red Cross/ARRL MOU to assist with the communication needs of the relief operation and reunification. From the ARRL operator team leads and the SM, we developed a second compromise with Fire Department Lt. Figueroa, and Chief Alberto Cruz by deploying an operator to the headquarters of the fire department in Juncos.

We established communications from the fire department headquarters to the JFO in San Juan. We received access for our operators to install a portable station at all fire houses in Puerto Rico, including Vieques and Culebra, supplying shelter, water, food, electricity, a place to sleep, and help to install the antennas, as well runners who could meet the mayors and government representatives near the fire house.

Also, we collected emergency requisitions for transfer by the ARRL operators directly to ARC DRO K1M and FEMA JFO W1M. Moreover, a fire house is a safe haven; we were able to deploy teams of one at each fire house required per impacted region. Taking advantage of the fire houses resources, we were able to manage the operators more efficiently from teams

Val Hotzfeld, NV9L, and Mike Corey, KI1U, staff the station at the Joint Field Office in San Juan. [Oscar Resto, KP4RF, photo]

of two to a single operator, keeping the ARRL operators safe.

At the east coast, my assistant SM Jose (Otis) Vicens, NP4G, covered the region of Humacao to Fajardo, and Section Emergency Coordinator Juan Sepulveda, KP3CR, covered from Lares toward the west coast. They both covered the traffic at the most needed times for emergency communications.

Many of the local radio amateurs had limited communication participation during the emergency since they were directly impacted by the loss of their station and the lack of electrical energy, as well as an emotional impact. The ARRL relief radio operators were located on the main stations including the Red Cross DRO at the Angel Ramos Building, the headquarters of the Disaster Relief Operation, and FEMA (JFO) at the Pedro Roselló Convention Center. Emergency portable stations were installed in many municipalities such as Vieques and Culebra, Mayaguez, Guayama, Juncos, Fajardo, Bayamon, Caguas, Lares, Castañer, Ponce, Guajataca and many others. The amateur operators collected and retransmitted the requisitions and patient's orders by amateur radio frequencies via voice/CW and Winlink emails.

Few repeaters where utilized during the six-week relief efforts due to the extensive antenna/tower damage and the power outage. About 13 repeaters came back into service after minor antenna repairs and reconfiguring the electrical power to emergency generators lines. Linked repeaters administrated by Ramón E. Ramos, KP4DH, had coverage from the east to the west cost of Puerto Rico, allowing emergency portable stations to coordinate easily with the ARC DRO K1M and FEMA JFO W1M.

I cannot finish my report without personally acknowledging the American Red Cross and the ARRL for their unconditional support and effort during post-Hurricane Irma/Maria times in the Caribbean Region. I cannot forget the valuable help of the 24 operators who traveled to Puerto Rico and gave us their unconditional support during our catastrophe. This team and the four team-leads helped 24/7 for the benefit and relief of our citizens in distress. These amateur operators joined our emergency communication effort at the most needed time in modern history.

I also cannot forget FEMA and Homeland Security (especially EFS-2), the SHARES team, and the Winlink organization who supported us during this period. Also I have to recognize the sacrifices of the 131 local amateur radio operators who, even though they were directly impacted by a Category 5 hurricane, devoted their time with limited resources to maintain the local communications across Puerto Rico. My personal gratitude to all of them!

I can share thousands of stories and anecdotes, the ones I know and many that I never heard about it. This is a short report of an incredible ham radio event — amateur radio is the one that works when everything else fails. This is a fact! The FCC, FEMA, Homeland Security Emergency Management, and the Red Cross all recognized what the volunteer radio amateurs did during Hurricane Maria during the effort for Puerto Rico.

ICS-205 INCIDENT RADIO COMMUNICATIONS PLAN		Incident Name: Hurricane Maria				Date/Time Prepared: 9 OCT 2017			Operational Period Date/Time: 1600 28 SEP 2017 Z	
Ch #	Function	Channel Name/Trunked Radio System Talkgroup	Assignment	RX Freq N or W	RX Tone/NAC	TX Freq N or W	Tx Tone/NAC	Mode A, D or M	Remarks	
1	Amtr Comms	KP4DH		146.67	94.8	146.07	94.8	A	North Central; Linked	
2	Amtr Comms	KP4IA		145.37	No PL	144.77	No PL	A	Aguas Buenas; PR wide	
3	Amtr Comms	Local Repeater		147.31	88.5	147.91	88.5	A	East PR to St. John	
4	Amtr Comms	Local hams		146.59	No PL	146.59	No PL	A	Power company and local amateurs	
5	Amtr Comms	ARRL Deployment		146.55	No PL	146.55	No PL	A	Inter-team comms	
6	Amtr Comms	SATERN		14.265	No PL	14.265	No PL	A	Assistance passing traffic to CONUS	
7	Amtr Comms	NVIS		7.085 LSB	No PL	7.085	No PL	A	ARRL HF comms	
8	Amtr Comms	EOC/COE		5.235 USB	No PL	5.235 USB	No PL	A	Primary	
9	Amtr Comms	EOC/COE		7.360 USB	No PL	7.360 USB	No PL	A	Secondary	
10	Amtr Comms	Local Repeater		147.05	127.3	147.65	127.3	A	South and North	
11	Amtr Comms	Local Repeater		147.29	123.0	147.89	123.0	A	West	
12	Amtr Comms	Local Repeater		147.23	123.0	147.83	123	A	West	
13	Amtr Comms	Local Repeater		147.31	88.5	147.91	88.5	A	East	
14	Amtr Comms	Local Repeater		147.25	88.5	146.65	88.5	A	East	
15	Amtr Comms	Local Repeater		147.07	146.2	147.670	146.2	A	Southeast and West	
16	Amtr Comms	Local Repeater		145.25	88.5	144.65	88.5	A	Linked to 147.31	
17	Amtr Comms	Local Repeater		447.80	136.5	442.80	136.5	A	Arecibo	
18										
19										

ICS-205 form showing frequencies in use for Maria response. [Courtesy Oscar Resto, KP4RF]

REFERENCES

[1]Alexander Hamilton, "A Few of Hamilton's Letters" (1772), The MacMillan Company, London, 1903.

[2]R. Knabb, J. Rhome, D. Brown, "Tropical Cyclone Report, Hurricane Katrina," National Hurricane Center, Dec 20, 2005.

[3]National Hurricane Center website: **www.nhc.noaa.gov**

[4]National Hurricane Center and Pacific Hurricane Center

[5]WX4NHC Amateur Radio Station at the National Hurricane Center: **w4ehw.fiu.edu**

[6]More information about the National Hurricane Center may be found at **www.nhc.noaa.gov** and in the document "The National Hurricane Center-Past, Present, and Future" available at **www.nhc.noaa.gov/pdf/NHC_Past_Present_Future_1990.pdf.**

[7]For more information about Hurricane Betsy, see **www.weather.gov/mfl/betsy.**

[8]Hurricane Watch Net website: **www.hwn.org**

[9]For more information about Hurricane Mitch, see **www.nhc.noaa.gov/data/tcr/AL131998_Mitch.pdf**

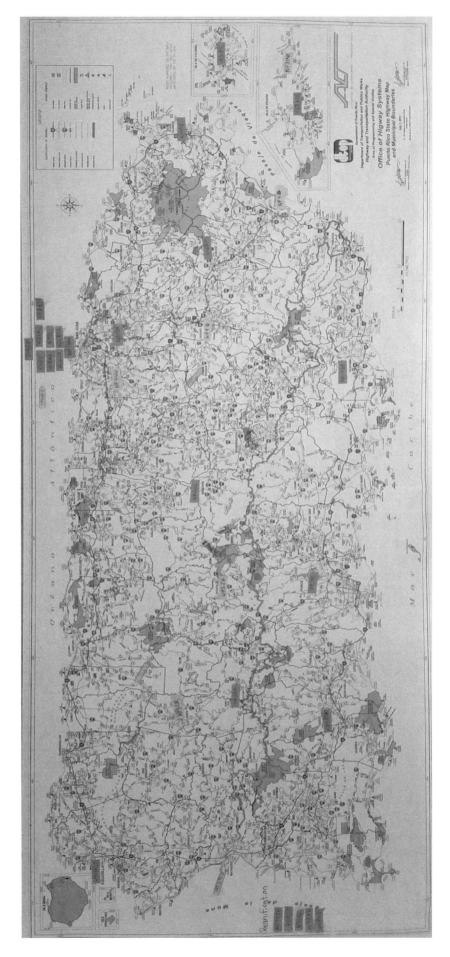

This map shows locations where amateur radio operator volunteers covered the Power Authority (AEE) Technical Plaza facilities. [Courtesy Oscar Resto, KP4RF]

Storm Spotter Activation

Ask any veteran storm spotter and they will tell you that there is more to storm spotting than simply responding to severe weather. The storm spotter first has to be prepared for severe weather and this happens long before the storm arrives or even before severe weather threatens. We have discussed some of the ways to prepare: training, equipment, safety awareness. During SKYWARN activation we often face challenges and issues that we normally would not face. And often, during activation, we are pulling double duty with an ARES or public service group. But once the severe weather has passed, our duties as storm spotters are not necessarily over. There are post-storm damage reports to be made, after action reports, ARES field organization reports, debriefings and possibly more work to be done assisting in the aftermath.

There is a lot that goes into being a storm spotter and probably even more as an amateur radio storm spotter. In this chapter we will cover some of the aspects of storm spotter activity before, during and after activation. Instead of going step-by-step through an activation process we will cover some key elements to storm spotter activation. Generally, these are the elements that can be applied to almost all amateur radio storm spotter groups.

We will start with activity that occurs before storm spotters are activated. We will cover getting ourselves ready for activation. From there we will walk through the days leading up to a storm: making sure equipment is in order, following forecasts and hazardous weather outlooks, communication prior to activation and understanding the roles we may play once activated.

Then we can look at issues that occur during activation. Storm spotter activity must respond and adapt to changing weather conditions, so we won't get overly specific about what happens during activation. Instead we will look at some critical issues that storm spotters may likely face: clear defined roles, storm spotter and SKYWARN nets, double duty between SKYWARN and ARES, lessons we can take from emergency management, and handling problems with storm spotters.

In the post-storm phase, we will look at how storm spotters can follow up on their activity. We will look at gathering storm damage reports, writing after action reports, submitting a public service activity report.

THE PRE-STORM PHASE

Get Into a Daily Habit

The first thing we need to do as volunteer storm spotters is to get into a daily habit of looking at forecasts and weather conditions. Many days may start out with sunshine and clear skies or no signs of severe weather, but over the course of the day may change to stormy conditions. We should monitor forecasts on a daily basis so that we aren't caught off guard. So, where do we get the information and what do we look for?

The best starting point is to go to the local NWS website and access the seven-day forecast for your area. This will give a sense of what to expect in the short term. If there is no indication of severe weather then we may not need to check in again during the course of the day. If there is an indication of severe weather, then investigate the forecast further. There are two sources of information that we can use to get more information about the forecast.

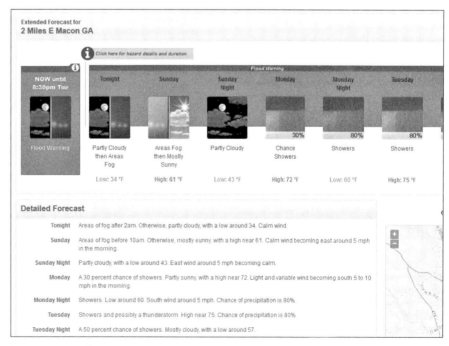

Extended local forecast. [National Weather Service]

First is the hazardous weather outlook (HWO). The HWO gives additional information on severe weather and the area it is expected to impact. It can cover a timeframe from the day of the weather event up to a week ahead. It also will let us know if storm spotter activation is expected or if specific reports are needed.

Two other sources to turn to when the forecast calls for severe weather are the Storm Prediction Center (SPC) and the National Hurricane Center (NHC). Both centers are part of the NWS. The SPC has forecast tools such as upper air maps, soundings analysis, mesoanalysis, and fire weather composite maps that we can look at and get a better idea of what to expect in the short term. The NHC provides information on tropical storms that are active in the Atlantic and Eastern Pacific. While tropical storms can be difficult to predict we can look at maps showing possible tracks over a three to five day period and whether the storm is likely to intensify or weaken.

There are, of course, other sources for weather forecasts. Resources such as television and radio weather broadcasts, conventional media sources, weather websites such as AccuWeather and The Weather Channel all provide good short-term forecasts. However, we go about getting the short term forecast each day, it is a good daily habit to develop. Taking a look at the NWS seven-day forecast and perhaps following it up with a look at the SPC or NHC information takes only a few minutes and gives the storm spotter valuable insight on what may lie ahead in their area.

There is one thing about forecasts we must keep in mind. If we look at the local NWS site or a similar forecast resource, we notice that it will generally provide a forecast for seven days. Weather can be difficult to predict even day-to-day let alone days in advance. Forecast technology has made it possible to predict weather several days in advance — but not with 100% accuracy. Most of us can recall times when severe weather struck with no warning. Many weather sources have available 10-day forecasts. We must remember that the further out the forecast goes the less accurate it will be.

Apps and Online Resources

Many storm spotters use mobile apps and online resources to monitor weather conditions either from a computer or smartphone. These are great tools to use to see what is heading our way, find current conditions, view watches and warnings, and integrate data such as APRS or CWOP. However, since most days we are not going to encounter severe weather, we may go weeks at a time without having to check in on these resources. Since these apps and sites are important to the storm spotter, we cannot afford to become rusty in using them. We need to look at it the same way we look at the radios we use. When we get on the air and make contacts we are, sometimes unknowingly, developing the skills we will use during severe weather events or in an emergency or disaster.

When severe weather isn't expected or present, we can take blue sky days to explore the apps and sites thoroughly. Go to radar sites that are indicating severe weather and try out different features and practice radar interpretation. The Storm Prediction Center offers a great snapshot of national weather, check there for places where severe weather may be present. This is also a great time to make sure your apps have been updated to the latest version, if your phone doesn't automatically update.

Vehicles

Proper vehicle maintenance is important at all times. For storm spotters there are some things that should be checked before activation. Make sure windshield wipers and tires are in good condition, top off the fuel tank and double check all emergency supplies (first aid kit, fire extinguisher, blanket, jumper cables, and so on). Make sure all communications equipment in the vehicle is working properly. Also make sure all lights are working properly: headlights, marker lights, turn signals, brake lights and interior lights. And make sure all routine maintenance is up to date.

Mobile storm spotting can be quite risky, more so in a vehicle that is not up to the task. By including a vehicle inspection in your pre-activation plan you are taking steps to stay safer while on the road in inclement weather.

The Storm Kit

One item the mobile storm spotter can have ready ahead of time is a "storm kit." Most of us are familiar with go-kits designed for either short term (less than three days) or long term (more than three days) use. The storm kit is similar in that it is a grab-and-go kit, but serves a different purpose. It is designed to have the tools a storm spotter needs ready to go and small enough to keep in the vehicle with them.

What do we put in a storm kit? Much like what to put in the go-kit, the answer depends on the individual and what they may respond to as a storm spotter. So let's take a look at a sample storm kit. Since it is going in the vehicle with us we don't want it taking up too much space or make it difficult to find what we need in a hurry. Some items like batteries, repeater directories, and spare parts for the handheld will be left to the go-kit inventory.

One item that we might keep in the kit is a small camera. As we discussed in the equipment chapter, a small digital camera can be a handy tool for the storm spotter. And many of today's cameras also have the added feature of capturing video. Even if your cell phone has the ability to capture good quality still and video images, keep a regular camera in the bag as a backup. A video camera is another option here. Many can capture still images as well as video at high resolution and aren't much larger or heavier than a still camera. The thing to keep in mind though is that these items rely on batteries. You will need to make sure that batteries are charged and ready to go when the kit is not in use. Consider purchasing an external, high capacity battery to charge your phone and/or camera in the field. Your local NWS office is always very appreciative when they receive images or video of severe weather phenomena that they can use in future education and training.

Smartphones today include fairly good GPS applications with maps that are regularly updated. There are a couple things to keep in mind if you plan on using your phone for GPS. First is that the GPS applications typically used on phones place a high demand on the battery. Make sure you have a good external power source. Second, most GPS apps require connection to a network to retrieve maps. Applications such as the app *Offline Maps & Navigation* or *Google Maps* allow you to download maps to use when a network connection is unavailable. Leave the road atlas for vacation and take along a topographic map of your area or a state atlas.

A handheld weather device such as an anemometer/thermometer would also be useful to keep in the bag. Having this available while storm spotting will allow you to report fairly accurate wind speeds and take out the guesswork. And additionally, it will help prevent reports of wind speeds that do not meet the reportable criteria. But like the cameras, an anemometer/thermometer may use batteries that require attention when the kit is not in use.

You should also have a blanket in your kit. During winter weather it can be used to keep warm if your vehicle gets stuck or dies and in warmer weather it can be used for protection from hail or debris from high wind. Remember to keep it inside the vehicle, not in the trunk.

And a final item is the paperwork. Most of what you need can be kept in a three-ring binder in your kit. At the least it should include notepaper, SKYWARN guide, your local SKYWARN or storm spotting manual, net and repeater information, NOAA Weather Radio frequencies, important phone numbers, and a map of your local NWS coverage area. The key is to make sure it is well organized so it is easy to find the information you need. Include with this pen, pencils and a marker. You can also make use of different types of cloud storage systems, such as Dropbox or Google Docs, to back up all your important paperwork.

There are other items that you will use while spotting that may already be in your vehicle: cell phone charger, radios, and perhaps GPS. The idea for the kit is to put those items you will need while mobile storm spotting in a grab-and-go bag. Imagine a carpenter arriving at a job site without a hammer. You don't want to respond to severe weather without all the needed tools.

Making sure you have what you need is not just for the mobile storm spotter. If you operate as a net control station or storm spot from a fixed location you want to keep certain items on hand. Make sure you have maps of the area. Computer programs such as Google Earth can provide great maps, but if the power goes out you might want to have paper maps as a backup. Also don't forget to keep water and snacks on hand. Net control stations may spend many hours on the air before, during and after the storm. And have your SKYWARN manual, net control procedures, spotter roster and storm spotter guides handy.

Make SKYWARN Part of Your Plan

One of the duties of your local Emergency Coordinator is to "Develop detailed local operational plans with 'served' agency officials in your jurisdiction". When a disaster strikes, we do not respond in an uncoordinated, unplanned,

```
                    Special Weather Statement
Special Weather Statement
National Weather Service Peachtree City GA
321 PM EST Sat Feb 8 2020

GAZ001>009-011>016-019>025-027-030>039-041>062-066>076-078-079-
081>084-091400-
Dade-Walker-Catoosa-Whitfield-Murray-Fannin-Gilmer-Union-Towns-
Chattooga-Gordon-Pickens-Dawson-Lumpkin-White-Floyd-Bartow-
Cherokee-Forsyth-Hall-Banks-Jackson-Madison-Polk-Paulding-Cobb-
North Fulton-Gwinnett-Barrow-Clarke-Oconee-Oglethorpe-Wilkes-
Haralson-Carroll-Douglas-South Fulton-DeKalb-Rockdale-Walton-
Newton-Morgan-Greene-Taliaferro-Heard-Coweta-Fayette-Clayton-
Spalding-Henry-Butts-Jasper-Putnam-Hancock-Warren-Troup-
Meriwether-Pike-Upson-Lamar-Monroe-Jones-Baldwin-Washington-
Glascock-Jefferson-Harris-Talbot-Crawford-Bibb-Twiggs-Wilkinson-
321 PM EST Sat Feb 8 2020

...BLACK ICE POSSIBLE DUE TO REFREEZING OF SNOW MELT TONIGHT OVER
NORTH GEORGIA...

...PATCHY FREEZING FOG POSSIBLE TONIGHT OVER MUCH OF NORTH AND
MIDDLE GEORGIA...

A fairly large area of 2 to 6 inches of snow fell over parts of
north Georgia today. Temperatures have risen above freezing in
most locations, but are expected to slowly drop back down to near
or below freezing tonight. Any water from snow that melts is
expected to refreeze in spots creating slick conditions. The
greatest threat for black ice from refreezing snow melt will be
over northeast Georgia north of a line from Calhoun to Canton to
Commerce.

Calm winds and some clearing skies will also allow for some
patchy fog develop over parts of north and central Georgia
tonight. Where temperatures drop to near or below freezing, there
is a chance that some freezing fog could develop which could cause
roads, bridges and overpasses to become slick. Residents over much
of north and central Georgia north of a La Grange to Macon to
Jefferson line should monitor the latest conditions before driving
tonight and Sunday morning.

Temperatures are expected to quickly warm above freezing after 10
AM Sunday morning, ending the threat for black ice and freezing
fog.

$$
```

The Special Weather Statement may be issued by the NWS for events that do not require an advisory or warning. [National Weather Service]

haphazard manner. If we did, the response could end up being just as a big a disaster as the disaster itself. Instead, we develop operations plans long before an emergency occurs. This is a team effort among all involved: amateur radio, emergency management, local public safety officials. Basically, it is a plan on how amateur radio plays a part and in what way during and following a disaster. Our involvement in SKYWARN can be included in emergency planning.

The Emergency Response Plan (ERP) is a document developed by counties, cities, and other local jurisdictions that describes "how citizens and property will be protected in a disaster or emergency."[1] The production of an ERP is coordinated by local emergency management, but all involved parties have input in the process of its development, and this includes amateur radio. The ERP document contains introductory information, the basic plan, annexes, and appendices. The annexes and appendices are usually where our role is found. The annexes explain how the community will carry out broad functions during an emergency. An appendix is a supplement to an annex that provides information on how to carry out a specific function in the face of a specific hazard.

One annex that is generally always

found in an ERP is one that covers warnings. This annex establishes a warning system within the jurisdiction that is capable of delivering adequate and timely information to officials and the public in the event of a threatened disaster. If we think about the systems in place capable of providing warnings to officials and the general public, we realize that there are more than one: the news media, the NWS, National Warning System (NAWAS), and warning devices such as tornado sirens and public alert systems. SKYWARN is a component within the watch and warning system. SKYWARN often relays reports to the NWS via a local emergency operations center or emergency manager. This provides a means to alert local officials. Some members of the public monitor our activity on a scanner and can be alerted of an emergency that way, while others will receive warnings from the NWS or media that may be the result of information that we called in.

Chances are your area has an ERP. To find out if amateur radio and SKYWARN are included check with your local Emergency Coordinator or emergency management agency. If they are not part of the plan, it may present an opportunity to get amateur radio and SKYWARN included. ERPs are not stagnant documents. There is a life cycle to ERPs. The first step is hazard analysis, then ERP development, then testing the plan, and finally maintenance and revision. SKYWARN is an important part of a community's emergency plan. If your group is not part of it, it should be.

Leading Up to Activation

We've looked at a few things that storm spotters can do to prepare themselves for SKYWARN activation. Now let's look at what we need to do in the days leading up to activation. Of course, there are times when SKYWARN activation is needed with little advance warning. Most times, though, we will have some advance warning through hazardous weather outlooks (HWOs) from the NWS. Let's use as an example of a forecast that calls for a chance of strong thunderstorms later in the week.

The HWO is the key to staying informed on approaching severe weather. The HWO is issued by the NWS and contains information on potential severe weather threats. Typically, it will contain date and time issued, affected areas, timeline of potential threats, and information

```
CTZ002>004-MAZ005>007-011>017-RIZ001>005-080930-
HARTFORD CT-TOLLAND CT-WINDHAM CT-CENTRAL MIDDLESEX MA-
WESTERN ESSEX MA-EASTERN ESSEX MA-EASTERN HAMPDEN MA-
SOUTHERN WORCESTER MA-WESTERN NORFOLK MA-SOUTHEAST MIDDLESEX MA-
SUFFOLK MA-EASTERN NORFOLK MA-NORTHERN BRISTOL MA-
NORTHWEST PROVIDENCE RI-SOUTHEAST PROVIDENCE RI-WESTERN KENT RI-
EASTERN KENT RI-BRISTOL RI-
422 AM EST FRI FEB 7 2020

...WIND ADVISORY IN EFFECT FROM 1 PM THIS AFTERNOON TO 10 PM EST
THIS EVENING...

THIS HAZARDOUS WEATHER OUTLOOK IS FOR NORTHERN CONNECTICUT, CENTRAL
MASSACHUSETTS, EASTERN MASSACHUSETTS, NORTHEASTERN MASSACHUSETTS,
SOUTHEASTERN MASSACHUSETTS, WESTERN MASSACHUSETTS AND NORTHERN
RHODE ISLAND.

DAY ONE
TODAY AND TONIGHT.

PLEASE LISTEN TO NOAA WEATHER RADIO OR GO TO WEATHER.GOV ON THE
INTERNET FOR MORE INFORMATION ABOUT THE FOLLOWING HAZARDS.

   WIND ADVISORY.

DAYS TWO THROUGH SEVEN
SATURDAY THROUGH THURSDAY.

NO HAZARDOUS WEATHER IS EXPECTED AT THIS TIME.

SPOTTER INFORMATION STATEMENT

SPOTTER ACTIVATION IS NOT EXPECTED AT THIS TIME.
```

Hazardous Weather Outlook text. This is a good source to determine if spotter activation may be needed in the near term. [National Weather Service]

on storm spotter activation. The HWO is issued by the local NWS office. As weather conditions change the HWO may be updated with new information. Unlike watches, warnings and advisories, though, there is no announcement made if an HWO is updated or cancelled so you have to check back regularly to keep up with changes. So how do we get this information? How do we incorporate it into local storm spotter planning?

We first start with our daily habit of checking the forecast. If you use a source other than your local NWS office website you may miss an issued HWO, so again it is best to go to the NWS for forecast information, or if other sources say there is a chance for severe weather. Now let's go to the example of the thunderstorm forecast.

First, we are going to go to our local NWS website and select our area by clicking on our county on the map or by entering our zip code in the search bar. Each NWS site may be a little different, but it shouldn't be too difficult to locate your local forecast information.

At the top of the screen we see the seven-day forecast and below that any watches, warnings, advisories, or hazardous weather outlooks.

We can see an HWO has been issued and it indicates that we can expect severe weather in the coming days. SKYWARN spotter assistance may be needed.

Once we have this information, we can pass the word to local storm spotters. This can be done by email, on a local net, with text messages,-or via phone calls. With this advance notice spotters can start making preparations for activation. After the initial HWO we need to make sure to check back regularly for updates.

Many local NWS offices utilize conference calling or NWS Chat to keep partners such as public safety officials, the news media and others updated when severe weather is expected within a 24 hour window. SKYWARN storm spotters may be able to utilize this. Check with the Warning Coordination Meteorologist at your local NWS office to see if you can participate.

So, let's say we've followed the HWOs that have been issued over the course of two to three days and now it is looking like there is a risk of severe thunderstorms, hail, flooding and possibly tornadoes in our area. The local NWS office informs their partners via conference calls and NWS Chat. During the call, information is given about the approaching severe weather, risk potential for locations within the NWS office's area, and perhaps storm spotter information. A website may also be utilized to present graphical information. The call may conclude with a chance for participants to ask questions of the forecaster conducting the conference call.

Depending on local activation procedures this may be a good point to get in touch with other groups and agencies we may work with during activation. We may also want to determine a net control schedule or rotation, tally how many fixed and mobile spotters we may have and double check radios, the net control station, vehicles, and so on. These steps will help make certain that we are ready for storm spotter activation.

In the hours ahead of the approaching severe weather we will want to start keeping an eye on radar. Weather systems can change course, although generally not drastically. A storm system may appear to be heading our way. By monitoring it we can easily notice trends: gradual change in course, strengthening or weakening, new cells developing. We can read storm reports from areas it has impacted. This information will help us make the decisions leading to local activation of storm spotters.

At this point we must look at who makes the call for activation. The local NWS office will make it known that storm spotter assistance/activation may be needed, but generally they won't make the call to activate storm spotters in a particular area. They may request a SKYWARN net be activated; this would definitely be the case when primary SKYWARN net control is located at the NWS forecast office. The activation of local spotters is primarily a local responsibility. This can be made by the local emergency management director, Emergency Coordinator, local SKYWARN coordinator, ARES leader, or whoever may be designated to make that call. In some cases, a formal activation of spotters at the local level may not happen. This does not mean an individual spotter cannot be active and submit reports. Volunteer storm spotters, whether as a group or individually, may be self-activated.

Let's take a look at two types of local spotter activation: group activation and individual activation.

STORM SPOTTER ACTIVATION

The local storm spotter group may be formed in a variety of ways. It can be a formal SKYWARN group consisting of amateur radio and non-amateur radio members. It can be organized through local emergency management. Or it can be a loose association of trained spotters. There are three key characteristics regardless of how the group is formed: they are trained storm spotters through the SKYWARN program, they are organized in some way and they have the ability to communicate with each other and with the local NWS office. For our discussion we are going to look at a local group of amateur radio operators serving as SKYWARN volunteer storm spotters.

The first step in activation, a request for assistance, comes from the local NWS office. Before severe weather arrives the local NWS office, if equipped, may activate their amateur radio station to serve as the primary net control station for SKYWARN. At the local level the decision is made prior to the arrival of expected severe weather to activate the local SKYWARN net and place it in standby mode. So, two different nets may exist during a severe weather event: the NWS SKYWARN net may be active, taking reports from stations throughout the NWS CWA (County Warning Area), and the local net may be taking reports from local storm spotters to be relayed to the local NWS office. EchoLink/IRLP is an excellent resource to link repeaters together so the NWS amateur radio station can go direct with local SKYWARN nets.

At the NWS level the net control station may be manned whenever there is a severe weather watch in the CWA or when weather conditions are such that activation may be likely. At the local level, placing the net and spotters in standby mode should be done in a similar fashion, with enough lead time to allow storm spotters to make preparations.

At this point we would begin making calls by radio, telephone, email, social media, or other means to local storm spotters to inform them of a pending activation. We would also get in touch with local served agencies such as public safety or emergency management to notify them of our status. This is also a good point to make sure that our lines of communication with the NWS office are working. Remember to have backup lines of communication — don't rely on only one method of staying in touch with NWS. During this standby mode we can begin to determine how many mobile and home-based spotters we have, including relay stations, net control stations, EOCs, and so on, for when we go fully active with the SKYWARN net. Don't forget during standby mode to continue to monitor NWS announcements, HWOs, radar and

NWS Blacksburg, Virginia, WFO SKYWARN Volunteer Carter Craigie, N3AO. [Kay Craigie, N3KN, photo]

conditions in areas already impacted by the storm. Critical information should be relayed to stations that have checked into the net during the standby period.

A common problem SKYWARN groups encounter in the period of time leading up to activation is spreading the word to members that activation is imminent. There are several reasons for this problem: time of day (early morning, during work day), members not near a radio, not enough SKYWARN members, repeater issues, and other factors. Today we have more means of communication than ever before: telephone, email, text messaging, apps such as WhatsApp and GroupMe, and, as amateur radio operators, radio. When we need to get the word out, we need to utilize all possible means.

Local SKYWARN coordinators can help alleviate some of these problems by taking several steps long before severe weather strikes. First, keep a current roster of all local storm spotters. This should include name, call sign, amateur radio capabilities, email address, home and cell numbers. Also include best method for different times of the day. Contact by work phone or cell phone may be best during the day, radio in the evening, and home phone at night. Don't rely on one method. Local SKYWARN coordinators should also make sure that there are enough training opportunities throughout the year. More training means more and better trained spotters. And coordinators should have plans for what to do if repeaters fail. Always plan to have backups! Make sure to utilize group text/chat resources such as WhatsApp and GroupMe, as well as social media platforms such as Facebook or Twitter to communicate with local storm spotters.

Once the severe weather has entered our response area, which can be defined as city, county, metro area, or other jurisdiction, we can move the net into active mode. We may continue to get more storm spotters checking in, but the net is ready for business. Now let's take a look at the local SKYWARN net and the area wide SKYWARN net that may be operating from the NWS office. And we will take a look at the roles that play a part in the SKYWARN net and process of submitting storm reports.

The Wide Area SKYWARN Net

The wide area net, which may or may not be active during a local activation, is the amateur radio station that operates during a weather emergency from the local NWS office. It is staffed by local amateur radio operators who are not only trained SKYWARN storm spotters but often given additional training in net control operations and NWS procedures. The wide area net may be conducted from the NWS office or at a separate location where information is then relayed to the NWS office. The local NWS office staff, in coordination with the assisting amateur radio operators, determines the conditions and time frames necessary for SKYWARN net activation at the NWS office.

Utilizing Social Media to Network Your SKYWARN Team

The opinion on social media among radio amateurs is varied; some call it a threat, a nuisance, a resource, another tool in the tool box. Social media, particularly Facebook and Twitter, are useful tools for the local amateur radio SKYWARN team.

Facebook allows users to create interest groups. Facebook group pages allow for users to connect to one another, share announcements, post storm photos and video, and have ongoing discussion on blue sky days. The SKYWARN team at National Weather Service Boston/Norton — WX1BOX — has utilized Facebook groups quite effectively for many years. Group members can see social media reports that come from others in the group and peruse photos from previous weather events. Moderators can create group events for SKYWARN classes. And links to online resources can be quickly shared.

Likewise, the Boston team has also utilized Twitter for SKYWARN purposes. WX-1BOX primarily uses the Twitter feed to repost information related to current weather events. These "Tweets" can come from local SKYWARN spotters, the media, the public, public safety offices, or even the National Weather Service office. Twitter allows for posts up to 280 characters and sharing photos/videos. It also features "hashtags" that allow users to follow specific threads. For example, Twitter posts during hurricane Maria included #Maria.

Social media can be a powerful tool for your SKYWARN group, but you have to know how to use it effectively to get best results. Make sure you're posting regularly. Otherwise, members of the Facebook group or Twitter followers may not look to it for current information. Use the social media rule of threes — each week post one item that is of general interest (a weather related news item or safety tip), one item on what your group is doing (exercises and drills perhaps), and one item that others can do to support your mission (build a go-kit or take online training).

The WX1BOX Facebook page.

Since this net may cover a very wide area, the net most likely will not be conducted on a local repeater, unless of course the repeater is linked in some way, for example by EchoLink, D-STAR, or wide area repeater network. So to accommodate local SKYWARN nets and storm spotters across the CWA, HF frequencies may be used.

The purpose of the wide area net is to collect reports gathered either from local SKYWARN nets or SKYWARN storm spotters and relay them to the NWS forecasters. Having HF capability during events that impact a very wide area (more than one NWS CWA) also allows information to be passed on to other NWS field offices if needed. For example, a tornado touches down at the southern edge of a NWS CWA and is reported to a local net by a storm spotter. The local net can relay this information to their wide area SKYWARN net. Since it fell on the edge of the CWA it may be an issue of concern to the NWS office to the south. This information can be relayed from one wide area net control station to the other, assuming both NWS offices have amateur radio stations.

If your storm spotter group utilizes social media, such as Facebook or Twitter, you will also want to make sure traffic on those sites is monitored, vetted, and passed on to NWS. This is a good opportunity to include younger amateurs in storm spotting as their social media skills may be more advanced than some older radio amateurs; and veteran storm spotters can help them learn as reports are vetted.

The Local SKYWARN Net

Ideally the local SKYWARN net should be conducted like the wide area net. The local net can operate from just about any location, but typically it is operated from a home-based station or a station located in an EOC or similar facility. In some places net control may also be the link between storm spotters and local officials. The storm spotters relay reports via net control who then relays them to the NWS and local officials. And local officials, as well as the NWS, may use net control as a point of communications with local storm spotters. The critical part to local net control is that they need to be able to stay operational throughout the activation or have a plan in place to transfer net control duties to another station if they are forced to go off the air. Loss of power, telephone, or internet can seriously hinder net control duties. But we also must keep in mind that the weather itself may force net control to shut down. We don't want net control to stay on the air if it is not safe.

The amateur or amateurs responsible for net control operations, like their counterparts at the NWS office, have a tremendous responsibility on their hands. They should not only be trained SKYWARN storm spotters but should also have training in emergency communications, traffic handling and net control operations. During severe weather, net control can be quite busy and it can be very stressful. Net control is not only responsible for the general management of the net but also often has the task of relaying critical weather updates, knowing where their storm spotters are located and helping keep them out of harm's way, receiving and relaying weather reports to the NWS office or wide area net control, and serving as local amateur radio point of contact for NWS, emergency management, EC/DEC/SEC and anyone else who may be involved in severe weather response. The demands placed on net control stations can be great, but it can be a very rewarding experience.

Local net control can be greatly aided by the use of relay stations or assistant net control stations. These stations can be tasked with handling the relay of information to NWS, passing weather related information to severe weather nets in neighboring counties, or monitoring weather updates coming from the NWS. Having this kind of assistance in place can allow the net control station to focus on taking care of the storm spotters who are checking into the net. These assistants can be stations at locations different from net control, or there can be another amateur who works alongside the net control operator to handle these tasks.

During net activation the net control station must be sure to keep track of what is going on. Net activity can include incoming and outgoing traffic from fixed and mobile stations, the NWS, the wide area net, neighboring SKYWARN nets, as well as online and social media sources. If there are mobile spotters, net control will need to know where they are in relation to the storm and keep them posted with current weather information to keep them safe. Net control will want to be certain to keep careful record of all these activities. We have to go beyond pen and paper methods. A computerized log can be a tremendous asset. Along with activities, be sure to log dates, times and all stations involved in the activation.

Now that we understand the types of nets involved in a storm spotter activation, we must take a look at the role the storm spotter plays.

Role of the Storm Spotter

The storm spotter, the vital link between what is happening on the ground and the NWS, can operate in two different ways, mobile or fixed. Both of these have advantages and disadvantages. And during activation it is ideal to have plenty of spotters active from different vantage points. There are some things that spotters, whether fixed or mobile, must keep in mind during SKYWARN activation.

First is your number one priority, safety. Home-based spotters should make sure that they are safe to stay on the air. They must make sure there is proper grounding of all equipment in case of lightning strike. Care must be given to make sure the home or structure that they are in and all inside are safe from severe weather. They must keep alert for approaching severe weather and changing conditions.

Before mobile spotters head out for SKYWARN activation they should first make sure their family is safe and their home is prepared for severe weather. Do not head out until this is done. No spotter's safety should be compromised. Mobile storm spotters must stay extremely alert for changing conditions. Net control can help keep both spotters posted on important weather updates. If it is not safe to be on the road because of the weather don't go out. During some severe weather events a state of emergency or curfew may be declared, limiting who can be on the road. These must be followed. What a storm spotter does is a valuable service, but we are not above the law or safety rules.

Let's take a look at how home-based and mobile storm spotters function when SKYWARN is activated. We will specifically look at the purpose of storm spotters during activation, some operational guidelines, and the advantages and disadvantages of both.

Regardless of whether a storm spotter is operating from home, mobile, or any other location, their purpose is the same: to relay important weather conditions that meet set criteria to an established local reporting point (net control, emergency management, and so on) or directly to the NWS. The criterion that must be met is established by the NWS. In Chapter 5 we discussed different types of weather conditions and generally what the reportable criteria for each type are. For the purpose of this portion of activation discussion we are assuming our local spotters are operating through a local SKYWARN net. Since the net is active, it is far easier to allow net control to pass on reports than to attempt to do so yourself from home and especially mobile.

There are a couple of key things to remember about calling in reports. Only call in reports that meet the criteria. Do not call in a report that it is raining where you are (unless it meets the rainfall criteria to be considered heavy rain) or that there is lightning. Lightning is not reportable because it is so common and the NWS has access to lightning data. The spotter should carry their SKYWARN manual or a card/sheet with reportable criteria listed.

We also must keep in mind the purpose of the SKYWARN net: to collect severe weather reports from storm spotters. This net is not a round-table discussion or the place to generally talk about the weather. Radio traffic should be limited to weather reports, traffic related to safety, welfare and emergencies and traffic directed solely to net control. If you have to contact another spotter or a station not checked in to the net, first get net control's OK to move off frequency and second make the contact on a frequency not being used by the net. Once you are back on frequency let the net control station know. And, along similar lines, advise net control if you have to step away from the radio, will be temporarily off the air, or have to close down. If for some reason you are off the air suddenly, such as a power failure, make every possible attempt to let net control know, or tell another station who is checked into the net, so they know what happened and that you are OK. This is especially true if you are operating mobile and suddenly have to go off the air.

Home and mobile spotters should follow the directions of the net control station. Many times net control is acting on instructions from the NWS, local emergency management, or possibly local public safety. Just as much as spotters are the source of weather information, net control stations are the source for a wide range of information beneficial to spotters. Net control stations may pass along information on new approaching weather, requests from the NWS or emergency management, information on roads being closed, or if a state of emergency has been declared.

Home-based spotters often have the advantage that they can provide assistance to net control with other tasks. Home-based spotters are reporting what they observe from a fixed location. While doing this they may be able to assist net control by monitoring NOAA Weather Radio, maintaining contact with the net control station at the NWS WFO, relaying information to neighboring SKYWARN nets, or by monitoring radar. The amount of additional assistance the home-based spotter can offer will of course depend on their ability to multitask and only the individual spotter knows their own capability to do this. But remember to not take on so much that you neglect the first rule, safety.

One of the advantages of a mobile spotter is the ability to set up and observe weather conditions from a vantage point selected by the spotter. This can make it easier to spot specific weather events, follow a storm as it moves, and leave an area quickly if conditions become unsafe. The one thing that a spotter must keep in mind at all times while mobile is safety. We've covered safety issues specifically in an earlier chapter, but they bear repeating here. Rules of safety must be adhered to religiously in a mobile environment. The weather presents hazards directly with rain, wind, hail, tornadoes, and so on, and indirectly with hazardous road conditions, distracted drivers and limited visibility. In the car we have several sources of potential distractions that we must stay aware of such as a second spotter, radios, GPS, lights, or cell phones. Combine all of this with increased excitement levels and stress. Mobile storm spotting is not for everyone.

For those who do storm spot mobile, though, careful consideration must be paid to how we set up to observe severe weather. The first step is to find out where the storm is and the direction it is heading. Assuming it is a thunderstorm we will want to stay on the right rear flank of the storm. This gives us the best vantage point to observe severe weather if it develops and at the same time keeps us out of the storm. As the storm moves away from us we can change position to continue our observation.

When selecting a spot to observe from there are several factors to keep in mind. First, choose a location that gives you as much view of the sky as possible. Try to avoid any obstructions such as trees, buildings, or signs on any side of you. Second, have a planned escape route. Your escape route should be a roadway that is not likely to be made impassible by debris from the storm. Roadways with trees on either side or overhanging branches should be avoided altogether: stick with major roads such as highways. Third, make sure you are still within communication range. Make sure you can still access the repeater and have cell phone reception. Fourth, avoid being a traffic hazard. This goes back to safety. Do not set up on the shoulder of the road or in any way that may

The Importance of Being Honest

As storm spotters we are trained to report accurately. Trustworthy reports make volunteer storm spotters a valuable part of the warning system. As much as we try to relay the right information, sometimes we may be mistaken. An isolated, honest error is tolerable, but routinely inaccurate or misleading information brings bad consequences, including loss of credibility and the risk of conflicting information delaying an important warning or advisory.

Some individuals deliberately have made false or misleading reports. In 2007, the NWS office in Milwaukee, Wisconsin, received over 25 false reports from the same computer in one weekend through the WFO's web page for storm report submissions. The person, who appeared to have some knowledge of storm spotting and severe weather, falsely reported tornado damage and injuries. Local media interrupted broadcasts because of warnings being generated from the NWS office. Complicating the situation, legitimate severe weather was present in the area and a tornado warning had been issued based on storm spotter reports. The FBI investigated and since then, new measures have been put into place at the WFO to reduce false reports.

Unintentional false reports often result from inadequate training, poor communication and honest mistake, or a combination of factors. The consequences may be more training, better coordination, or perhaps restricting an individual from acting as a storm spotter.

Knowingly submitting a false storm report is against the law and is a violation of the False Statements Accountability Act of 1996 (18 USC §1001 — see Appendix 5). Deliberately violating this federal statute can result in an individual being fined up to $250,000 and/or being sentenced up to five years in prison.

present a hazard to you or other motorists. Next, respect private property. When you set up remember to do so from a location accessible to the public. What you do as a spotter is important but it does not give you license to be disrespectful to another person's property.

These steps to choosing a location for spotting are not merely suggestions but are all imperatives on the storm spotter that relate to storm spotting safety. Most of them can be addressed long before severe weather arrives. If you are dedicated to storm spotting mobile, find locations to spot from before severe weather strikes.

So now we have the actors in our storm spotter activation. The purpose of our response is to submit reports on severe weather. So what happens with these weather reports? Where do they go and how do they get used once they reach the NWS WFO?

Severe Weather Reports

Severe weather reports, they're the reason we serve as storm spotters. But if we were asked by John Q. Public, "What happens when you call in a report?", how would we answer? In this section we will look at this and also discuss the importance of our storm reports, the role they play in the integrated warning system, and the role they play in the larger severe weather team. We will then look at how the reports are actually received by the NWS. Finally, we will look at what the NWS does once they receive them: preliminary Local Storm Report (LSR) and Storm Data.

We know storm reports are important, but what do they actually provide? There are three things that storm reports provide to the NWS: ground truth information, situational awareness, assistance in verification efforts.

There are many attributes of a storm that the NWS can detect from their local field offices. Radar can be used to detect precipitation, tornadic signature (a hook echo), high winds (bow echo) and the probability of hail. Satellites can be used to observe cloud structure and water vapor. All of these things are occurring far above the ground, far above the people and property that weather affects. What is missing is observation at ground level, information on what is being experienced. Ground-based weather stations can provide a lot of information but they cannot tell you if there is a wall cloud or the diameter of hail. Storm spotters help provide ground truth information on what is happening. This information, combined with data from radar, satellites, weather stations and other sources, helps give meteorologists at the NWS a more complete picture.

This more complete picture is what the second element of storm reports is all about: situational awareness. Information is what provides situational awareness. The more information received the more aware we are. We can get a feel for what this is by looking at a 911 operator as an example. The 911 operator receives a call for a crime in progress. They gather information from the caller (ground truth report). They can access previous calls for service at that residence or business (gather additional information). They dispatch police to the scene. Situational awareness is the ability to take all of the incoming information and paint a picture of what is happening to those who need to make critical decisions on the course of action to take. For the forecasters at the local NWS office this kind of awareness helps in issuing warnings, updating forecasts and providing critical life and property saving information to the public.

The information received from storm spotters also assists in verification efforts. As discussed above, radar, satellites and weather stations provide valuable information but do not give the total picture. When information from these sources is combined with ground truth reports, meteorologists can better verify what is happening and what might happen. Storm spotter reports also play a part in post storm verification efforts. Storm spotters may submit information after the event that helps NWS staff verify what happened. For example, was the damage due to a microburst or a tornado?

A radar "hook echo," often a precursor of tornadic activity. [National Weather Service]

Integrated Public Alert and Warning System (IPAWS)

IPAWS is a national warning system for local alerting. The system supports sending alerts from local, state, tribal, territorial and federal officials during an emergency. These authenticated alerts provide the public with information on threats and hazards, along with relevant safety information. The alerts may arrive via radio and television via the Emergency Alert System, cellular phones via Wireless Emergency Alerts, NOAA All Hazards Weather Radio through the IPAWS-NOAA gateway, and internet applications.

NWS, as part of the federal system, may utilize IPAWS to disseminate alerts related to severe weather threats. Information used to generate these alerts comes from a wide range of sources, including storm spotter reports. Others on the team that play a role in gathering threat information used for generating alerts include public safety, meteorologists, storm chasers, media, and the public. IPAWS effectiveness comes from the team behind it that helps gather data and information that can be authenticated so timely alerts are issued.

We, as storm spotters, are part of a team. Each member of the team plays a key part in the process of forecast, detection, dissemination and public response. When we activate as storm spotters, we see this teamwork in action. We work with the local NWS office, emergency management, and perhaps public safety. This team functions because of its ability to communicate amongst individual members and ultimately between members and the local NWS office.

How Reports are Received

Now that we know the importance of reports and the part they can play in IPAWS, let's take a look at how reports are received. Storm reports don't just happen on their own; storm reports are received because of two factors: first are the relationships between members of the team and, second, the technical means that make it possible to submit the reports. If we think about the spotter training, we received from the NWS, we recall that there is an emphasis on teamwork and support. The support from the NWS helps keep the team running and effective. This support can be training (SKYWARN courses), answering questions and concerns from storm spotters, providing educational material, and keeping lines of communication with spotters going even when severe weather does not threaten. Storm reports are also aided by the relationships and collaborative efforts the NWS offices foster with local media and emergency management. We must remember storm reports may come from sources other than SKYWARN weather spotters. The storm reporting team consists of the NWS, local public safety officials, news media, storm spotters and emergency management.

As amateur radio operators we are well aware that there are many ways that messages can be relayed. Today we have more options than ever before: radio, cell phone, social media, internet, and so on. And of course, there are communication systems utilized by the NWS such as NAWAS. We have come a long way since the days of weather reports coming in via telegraph.

Internet Tools

Without a doubt the internet has proven to be a great tool for submitting weather reports. Over the last few years several different platforms have emerged as tools to submit storm reports. A couple online resources should be part of the storm spotter tool box.

Social media has become an important way for the NWS to disseminate products and information to the public. Every NWS forecast office has a Facebook page to post information and also to receive information such as storm reports. There is also a national NWS Facebook page where interesting/important information is posted nationally. Many offices also have Twitter accounts for disseminating and collecting information such as storm reports. The NWS also maintains official SKYWARN Facebook page and Twitter feed. Along with growth in internet communication, the NWS has also realized the value of mobile platforms. Mobile applications are available for cell phones and mobile internet, so storm reports may be submitted from just about any place that can access these services.

Finally, are online reporting forms. Many NWS offices have an online form where storm reports may be made. The form asks for details that fit reportable criteria for that office. Unlike NWS Chat, this is not a secured system. You do not have to be a registered user to submit a report. Typically, though, they will ask for identification and contact details to aid in verification. And of course, IP addresses can be traced when fake and malicious reports are provided.

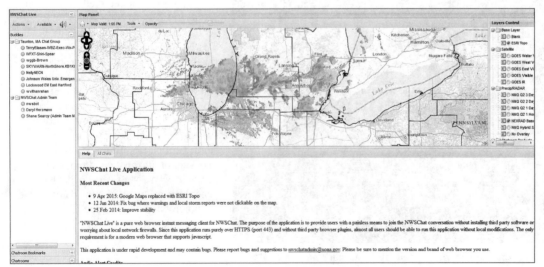

NWS Chat, a tool used for communication between NWS and many of their partners. [National Weather Service]

SKYWARN's Role

The method of reporting that we're most familiar with, SKYWARN, is another way information reaches the local NWS office. SKYWARN reports may come from amateur radio SKYWARN and ARES groups or members of the general public who have taken SKYWARN training. Reports from the SKYWARN community may come in via a net control station, radioed direct to the NWS amateur radio station, or by some other means such as social media, webform or NWS Chat. We can see SKYWARN as both a means and source of reports. Amateur radio stations located at NWS office offer another means of communication while non-amateur radio SKYWARN spotters are a source that utilize other means.

Automated Systems

Storm spotters know that what we report meets certain reportable criteria; for example a funnel cloud is reportable while lightning is not. There are times, though, that specific weather data is needed. The NWS has several methods for data to be reported automatically. These data can be a valuable part of the forecast and warning process.

One such weather data source is a mesonet. A mesonet is a network of automated weather stations that report mesoscale weather phenomena. Since the purpose is to observe mesoscale events, for example squall lines, mesonet stations are typically located close together and report fairly frequently, one to fifteen minutes.

Two other automated reporting systems that may be utilized are the Automated Surface Observing System (ASOS) and the Automated Weather Observing System (AWOS). ASOS stations, which are operated by the NWS, the Federal Aviation Administration and the Department of Defense, provide near real-time weather observation data to the NWS as well as the aviation, meteorological, hydrological and climatological research community. The information is updated each minute. Reports from ASOS stations typically occur each hour, but may be made more frequently if weather conditions exceed certain parameters. Data from ASOS stations include cloud height and amount to 12,000 feet, visibility to at least 10 statute miles, fog, haze, pressure, ambient and dew point temperature, wind direction, speed and character, precipitation accumulation and they may be set to look for changing conditions such as sudden changes in wind or pressure.

AWOS stations are operated primarily by the Federal Aviation Administration, not by the NWS or Department of Defense. These stations report weather data about every 20 minutes. The data reported are temperature and dew point, wind direction and speed, visibility, cloud coverage and ceiling and altimeter setting. Sensors may also be added that provide current weather conditions, detect freezing rain, and provide lightning data.

CoCoRaHS

Another voluntary observation network to be aware of is CoCoRaHS — the Community Collaborative Rain, Hail and Snow Network. CoCoRaHS is a unique, non-profit, community-based network of volunteers of all ages and backgrounds working together to measure and map precipitation (rain, hail and snow). By using low-cost measurement tools, stressing training and education, and utilizing an interactive website, this network provides high quality data for natural resource, education and research applications. National Oceanic and Atmospheric Administration (NOAA) and the National Science Foundation (NSF) are major sponsors of CoCoRaHS.

National Early Warning System

Early on in the Cold War, NAWAS, the National Warning System, was implemented as a means of warning of enemy attack or a missile strike on the US. Since the Cold War ended NAWAS has been primarily used for natural and technical hazards. NAWAS is a telephone system, essentially a party line that connects government users and allows them to communicate with each other. The system has a couple of built in safeguards that help to keep it on line and was recently upgraded by FEMA. It has built-in lightning protection and the lines avoid local telephone switches so they can stay on when circuits go down or are overloaded. Secondary users of the system include local emergency management, the NWS and local public safety answering points (PSAP). These users can stay in easy contact using the NAWAS line. Information on severe weather may be relayed to the NWS and the NWS uses the system to disseminate severe weather warnings.

Direct Solicitation

Another way the NWS can obtain storm reports is through direct solicitation. When the local NWS office puts out a statement containing weather information they may attach at the end of it a request for specific weather information and reports. For example, during a week where heavy rain is expected they may request that storm spotters and local law enforcement report any flooding. This typically may be found in issued hazardous weather outlooks. The NWS may also contact a local agency

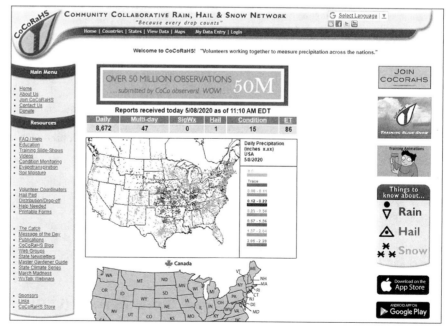

Website of CoCoRaHS — the Community Collaborative Rain, Hail and Snow Network.

directly or make a request through a SKYWARN net for specific information.

Home Weather Stations

Home weather stations also provide a source for storm reports. These weather stations can be used for reporting purposes in a couple of ways. If they can be connected to the internet they can be set up to automatically report data to the Citizen Weather Observer Program (CWOP). This information can then be accessed by the local NWS office, local public safety officials, emergency management, and storm spotters. However, the station does not have to be connected to the internet to be useful. Weather data may be monitored by the station owner and submitted to the local NWS office. For example, if the NWS office is seeking information on rainfall rates during a period of heavy precipitation the home weather station owner may monitor their station and make reports on hourly rainfall amount. And during hurricanes, home weather stations can be used to report weather data such as barometric pressure, wind direction and speed, and rainfall amounts.

Webcams

Webcams also provide a way the NWS can receive severe weather information. Webcams can serve many purposes. As we discussed in a previous chapter, WeatherBug uses webcams in their network of weather stations. Many TV stations and transportation departments have "tower cams" that give a view of metropolitan areas. These can all be used by NWS personnel as a way of seeing what is happening. Like a storm spotter they can provide an extra set of eyes to observe weather conditions.

Aviation and Marine Communities

The aviation and marine communities are another source of severe weather reports. Both have a vested interest in weather conditions and regularly monitor current conditions and trends. There are dedicated weather monitoring stations in each of these communities. We discussed earlier the ASOS and AWOS systems used in aviation. The marine community also has dedicated weather stations called weather buoys. These instruments collect weather data on the oceans of the world. Data collected includes air and water temperature, wave height, barometric pressure, and wind speed (sustained and gusts) and direction. Some buoys are moored while others drift. Data from these buoys can be monitored through several sources including NOAA's National Data Buoy Center, Weather Underground, and Stormpulse. Also private citizens in the marine community can contribute valuable reports. Boaters around the world can relay detailed information on storm conditions.

We discussed earlier that the NWS does not endorse storm chasing. Obviously this does not prevent storm chasing from happening. Storm chasers can provide a valuable service to the study of severe weather. Many colleges, universities, and research centers have active storm chase teams. And there are those who storm chase for personal or commercial interests. In either case storm chasers can provide local NWS offices with valuable storm reports. Storm chasers typically have significant meteorological knowledge and training. Many storm chasers also carry equipment to measure weather conditions, capture pictures and video, and for communications. All of this can make it possible for the storm chaser to quickly relay accurate, critical weather information to the NWS office.

Post-Event Reporting

Storm reports not only come in during the weather event, but also after the event. Following severe weather, information continues to come in about the storm's aftermath. Videos and still images provide details on damage from the storm. Media reports provide details on damage and how populations were impacted. And accumulated data may be submitted, such as total storm rainfall in a particular area. As storm spotters we must be aware that post-event reports are critical. They help meteorologists and hydrologists understand more fully what happened, aid in verification efforts, and make a full report on what happened possible.

Storm reports are absolutely essential in the IWS process. Because of this high level of importance, it is necessary to use every possible source to obtain severe weather information. No source can be overlooked. Because of the goal of safeguarding life and property all reliable information must be considered.

Local Storm Reports (LSR)

So now that we've looked at the variety of ways that the NWS receives storm reports we have to ask ourselves, "What do they do with them?" The first step of course is to determine the accuracy of an incoming report. Generally, each report is considered authentic until proven otherwise. Despite this benefit of the doubt, accuracy of the report and consistency must still be established. Once this is done the report can move on to the next step, the preliminary Local Storm Report (LSR).

The LSR product is issued by the local NWS office. The LSR provides the Storm Prediction Center (SPC) with information on hazardous weather events. The SPC uses the LSR in hourly reports, available to adjacent Weather Forecast Offices (WFOs) and partners such as media, emergency management, and storm spotters. The LSR gives another level of awareness of developing weather conditions.

The content of the LSR can include information on tornadoes, waterspouts, large hail, flooding, strong winds associated with thunderstorms or marine gusts, or just about any other type of severe weather event. Like SKYWARN storm reports the LSR contains weather information that meets or exceeds warning criteria. The LSR content is limited to a single report of an event, not multiple reports of the same event. Likewise other information may be omitted such as unconfirmed events and events containing partial information that may result in confusion. Because of the information contained in the LSRs they are issued as close to real time as possible. This means that the process of checking for accuracy and consistency must be done as quickly as possible.

An LSR can also contain information beyond a single event. During the event an LSR may contain information that summarizes events that have occurred or are occurring. After the weather event an LSR may be issued which summarizes weather activities in the CWA. Each NWS WFO may issue LSRs however they see fit. This can be done as an LSR for each report during the event and a summary LSR afterward. Some may compile a list of LSRs generated for each report received into a single report, while others may issue only summary LSRs. Storm reports received in the post event period can help in compiling post event and/or summary LSRs.

Once the event is over, the post-event reporting phase begins. In this phase additional information is gathered that will later help in compiling the storm data. During the weather event the reports coming in are much like information gathered at the scene of the crime, to use a law enforcement example. The information helps determine courses of action, in this case forecasts and warnings. In the post-event reporting phase, to continue the law enforcement analogy, it is much like investigators and detectives collecting additional information. This process helps fill in any blanks and verify events that did happen. Since one severe weather event can quickly follow another, time is of the essence. A second severe weather event can make this process difficult or even impossible.

So, where do the post-event reports come from? We've already emphasized that storm spotters can assist in the process by submitting information on storm damage. The NWS WFO may make some follow up calls to verify events and conduct field investigations. Storm spotters may be asked, for example, if they observed any indications of tornadic versus straight line wind damage. Public safety emergency operation centers may be called to see if they received any weather reports, since reports made to these centers may not be immediately reported to the NWS due to other emergencies. Rural networks can also provide valuable post-storm information. Many rural areas may be more affected by communication interruptions during severe weather. The post-storm phase may be their first chance to submit reports. And NWS personnel may also gather media accounts and video of storm events. All of this additional information may go into LSRs issued after the event. LSRs may be issued for up to seven days following an event.

Storm Data (SD)

Once the reporting and investigation is conducted at the local level the WFO begins the Storm Data (SD) process. The SD process takes information from local WFOs and presents it via a Storm Events Database and in a monthly publication that is distributed through the NOAA National Centers for Environmental Information (NCEI) in Asheville, North Carolina (formerly the National Climatic Data Center, NCDC). The SD report covers "severe and unusual weather" that occurs across the US. The SD report contains highly detailed information about a storm. Local WFOs have 60 days from the end of the month that the event occurred to complete and submit SD.

SD reports are accessible by going to the NCEI's website.[2] A search tool allows users to find reports by state, date of event, county of event and event type. The user may enter additional search criteria such as tornado size, hail diameter, wind speed, number of injuries or deaths, or monetary damage to property or crops. Reports are then listed by location or county, date, time, type of event, magnitude, injuries, deaths and damage to property and crops. You may be surprised how many reports are available. For May 2019, there were 265 reports for Ohio alone.

The reports that we make as storm spotters are part of a great flow of information about severe weather. Our reports and the reports from other sources are essential in the Integrated Warning System (IWS). The products that are generated from the reports are the Local Storm Reports, which provide valuable information to storm spotters and other NWS partners and Storm Data which provides detailed information about severe weather events. The latter helps researchers with climatology of hazardous weather and flood events. Many of us, as storm spotters, have seen and perhaps used information contained in LSRs. And at times a report we may have submitted may have caused an LSR to be issued. But our reports have a life span far beyond the LSR. They are a part of the process of keeping an archive on severe weather and contribute to a better understanding about weather.

ARES AND SKYWARN

Several times we've mentioned the dual response nature of ARES and SKYWARN. Let's take some time to discuss the relationship between the two programs.

ARES, the ARRL Amateur Radio Emergency Service, consists of licensed amateur radio operators who voluntarily register their skills and equipment for communications duty in the public service when an emergency or disaster strikes. The only requirements for membership are an amateur radio license and a desire to serve. Coordination of ARES falls under the ARRL and is supervised by ARRL HQ staff. The organizational divisions of ARES are national, section, district and local. These levels are overseen by the HQ staff, Section Emergency Coordinator, District Emergency Coordinator and Emergency Coordinator respectively.

Damage from Worcester, Massachusetts ice storm, December 2008. [Rob Macedo, KD1CY, photo]

ARES members may respond to disasters that can include anything from hurricanes to wildfires, search and rescue activities to power outages, tornadoes to hazardous material spills. ARES is an amateur radio all-hazards response mechanism.

SKYWARN is a separate program from ARES. It is administered by the National Weather Service and is coordinated by a meteorologist at the local NWS office. The local WFO SKYWARN coordinator generally oversees the local SKYWARN program in their CWA, sets up SKYWARN classes and may recruit some local coordinators to provide assistance. They also work with the amateur radio operators who run the amateur radio station at the local WFO. Unlike ARES, SKYWARN is not an all-hazards response mechanism: SKYWARN is limited to severe weather.

Many amateur radio operators are involved in one or both organizations. The amateur radio emergency response, regardless of disaster type, is handled through ARES. It is the responsibility of the Emergency Coordinator to oversee the local amateur radio response during an emergency or disaster. This includes severe weather. Because of this, in many places ARES and SKYWARN are often synonymous. When severe weather strikes, we go into an ARES/SKYWARN mode. Severe weather reports are handled through the SKYWARN system, while emergency communication assistance is handled through ARES. This would be the dual-response aspect of ARES and SKYWARN.

Dual response of ARES and SKYWARN includes some assumptions. First, both groups must exist within a local area. Second, dual response indicates shared membership — ARES members are SKYWARN members. Third, the relationship between the two is understood. Fourth, individuals who are participants of both programs have been through the appropriate training.

There are also times where we go in to a single response mode. For example, a weather event may not require emergency communication assistance. Many times we are called on for SKYWARN duty and nothing more. Other times an emergency communications need exists but there is no severe weather event that would require a SKYWARN response. A hazardous materials spill would be an example. While information on such an event might have to be relayed to the NWS, it does not meet the requirements for SKYWARN activation.

Participation in SKYWARN does not automatically include ARES membership and vice versa. SKYWARN requires very specific training for its members. And ARES includes training that SKYWARN does not, such as ICS and EC-001.

Working with Volunteers

We all realize that SKYWARN and ARES are volunteer programs. Volunteerism is the heart of being an amateur radio operator, particularly during times of emergency. We've discussed how we are organized during disasters either through SKYWARN or ARES. For those

Damage from the May 2007 EF-5 tornado in Greensburg, Kansas. [Keith Kaiser, WA0TJT, photo]

in volunteer leadership positions, such as local SKYWARN coordinators and ECs, it is valuable to have some understanding of how volunteer organizations work, their strengths and weaknesses and how to handle problems that may come up with volunteers. We can learn a lot about this from the emergency management community.

One of the first steps that we can take is in training. FEMA offers a course through their independent study division called "Developing and Managing Volunteers." While this course is intended for those in the emergency management field, it offers a great deal to us in managing volunteers. Local SKYWARN coordinators and ECs can learn a great deal about volunteer management, recruiting, and handling stress among volunteers.

The first step for the SKYWARN coordinator or EC is to be familiar with the relationship we have with the served agencies. In the case of storm spotting this will primarily be the NWS and emergency management. While volunteers provide served agencies with many benefits, there are also challenges in working with volunteers. These challenges may be real or perceived. We're pretty familiar with the benefits that volunteers provide: cost effective services, access to a broad range of expertise and experience, free up paid staff to focus on other tasks, serve as a link to the community.

So what are some of the challenges? Training volunteers takes time. Can you say you were ready for all types of severe weather after one SKYWARN training class? Volunteers are not permanent. There are some who feel that technically competent people do not volunteer or believe that volunteers lower professional standards. There are insurance and liability issues with volunteers. And there can be a feeling by paid staff that volunteers are competing with them.

While there are some things we cannot change, there are some things that we can affect. As a local storm spotter coordinator you can play an important role in promoting benefits and alleviating challenges. This is not done alone: remember we're in a relationship with a served agency. Working with the staff of your local NWS WFO and local emergency management office is important in developing a strong volunteer-served agency partnership.

Recruitment

Another issue in volunteer programs is recruitment. SKYWARN and ARES both have very basic membership requirements. NWS recommends spotters complete a SKYWARN class offered by the local NWS office, and they should have the ability to communicate reports to the local office. ARES requires an amateur radio license and a desire to serve during times of disaster. Does this mean that both organizations only draw those that are highly qualified, competent, dedicated emergency communicators that can handle the stress and demands found during times of disaster? No. Is there a mechanism in place to weed out volunteers who pose a liability to the organization? The only mechanism in place is found in federal regulations. It is a crime to submit false weather reports to the NWS (see Appendix 5) and as amateur radio operators we have Part 97 of the FCC rules. Does this pose a serious challenge to the local SKYWARN coordinator or EC? Yes. So how can a coordinator or EC develop the best local emergency communications group possible? Through these steps: recruit, train, evaluate, and recognize/retain.

SKYWARN and ARES both recruit members through a number of sources such as websites, clubs, net, classes, or pamphlets. At the local level the SKYWARN coordinator or EC can develop methods to recruit amateur radio operators to both of these organizations. The first step is to develop a way of presenting information to potential volunteers that explains what the expectations are. For our purposes we are going to focus on severe weather and storm spotter response. So, we will want to explain that there is a training requirement, the NWS basic SKYWARN class. We will explain the role we play — as storm spotters we provide reports on severe weather conditions and as amateur radio operators we support NWS communications through our net control stations. We need to explain that severe weather response is not for everyone. Severe weather can make for high stress and high-risk environments. We need storm spotters who are safe, responsible, reliable, trustworthy, and can handle a high stress situation. We also need to explain what equipment an amateur radio storm spotter needs. And we need to explain the relationship amateur radio has with the NWS and local emergency management. People are more likely to volunteer if they are presented with as much information as possible on what is expected of them. Once we have a plan in place on how to present the information we need to look at where we can find volunteers. Consider several possibilities: hamfests, club meetings, VE sessions, SKYWARN classes, and nets.

Training and Evaluation

Once we have volunteers we need to move to the next step, training. Obviously the first step for a storm spotter is SKYWARN training. We don't want to stop there, though. We need to emphasize other training possibilities such as ARRL's EC-001 and Department of Homeland Security AuxComm, Advanced SKYWARN, or FEMA courses. Refer to the Training chapter for ideas. We cannot force people to commit to training beyond what is required. Make every effort, though, to encourage people to take advantage of every training opportunity. Work with your local NWS office to bring regular SKYWARN classes to your area. Just because you've taken it once doesn't mean that you can't take it again. Have local SKYWARN meetings where information about severe weather can be presented. Have veteran SKYWARN storm spotters as guest speakers. Be creative and encourage a learning environment. It is very easy for us to get lax in this area when severe weather is not threatening or during seasons of regularly good weather.

Evaluation is another key to a good local volunteer program. Evaluation occurs after an event. Evaluation of individual performance is needed after a SKYWARN activation has ended and the final assessments are being done. This is what tells us if there is a problem, a training issue, or if a member of our group went above and beyond.

And finally, recognize and retain. If we don't recognize volunteers for their effort, we will be surprised how fast they leave the group. Human beings by their very nature expect some form of reward for their efforts. Volunteers are no different. We may not receive monetary compensation for what we do, but payment in the form of recognition can be just as valuable. The amount of time it takes to recognize a job well done and make that recognition known is negligible. Recognition can come from the local coordinator or EC. It can come from a district or section level, DEC/SEC/SM. The ARRL has a certificate that can be used for recognizing the efforts of an amateur radio operator (Certificate of Merit, found under Field Services Forms).

Handling Stress

When working with volunteers we also have to learn how to handle volunteer stress. Stress should be anticipated whenever we volunteer as amateur radio operators to assist during times of disaster. Severe weather and storm spotting are no different. There are ways though we can address and handle stress among our volunteers.

Before severe weather ever strikes we can address the issue of stress through training. There are classes available that teach stress management skills. These may not be available everywhere or may cost money. An approach we can take is by utilizing expertise found in our local public safety agencies. Invite a police officer, firefighter, medic, or 911 dispatcher to a storm spotter meeting and have them talk about how they handle stress on the job.

During a severe weather event ECs and SKYWARN coordinators should make sure volunteers are well matched to their task. If someone cannot handle a fast paced, high stress environment, then placing them in the emergency operations center may not be a good idea. Also make sure that volunteers get regular meals and breaks and that they are rotated out after a reasonable length shift. This is particularly for severe weather events of long duration such as floods, ice storms, or hurricanes. These are the events that most likely will be SKYWARN and ARES events.

After the storm make sure you talk with the volunteers and find out what kind of stress they were under. Was there some way that the stress could have been alleviated? Do they need additional help such as Critical Incident Stress Debriefing? Severe weather can result in high levels of stress and trauma. Imagine if a storm spotter who was operating mobile came home to find that their neighborhood was hit by a tornado. Storm spotters are not immune to the trauma of severe weather destruction.

Here are some additional tips to dealing with stress and stressful situations:

• Delegate some of your responsibilities to others. Take on those tasks that you can handle.

• Prioritize your actions, the most important and time sensitive ones come first.

• Do not take comments personally. Mentally translate personal attacks into constructive criticism and a signal that there may be an important need that is being overlooked.

• Take a few deep breaths and relax. Do this often, especially if you feel stress is increasing. Gather your thoughts and move on.

• Watch out for your own needs: food, rest, water, medical attention.

• Do not insist on working more that your assigned shift if others can take over. Get rest when you can so that you will be ready to handle your job more effectively later on.

• Take a moment to think before responding to a stress causing challenge. If needed, tell them you will be back to them in a few minutes.

• If you are losing control of a situation, bring someone else in to assist or notify a superior. Do not let a problem get out of hand before asking for help.

• Keep an eye on other team members and help them reduce stress where possible.

Other Challenges

Imagine a SKYWARN activation for a severe thunderstorm. The NWS has indicated that there is potential for strong winds, large hail and possibly a tornado. We activate our SKYWARN net. There is a net control station, a relay station, a dozen or so home-based spotters and about half a dozen mobile spotters. Our area of responsibility is County X. Net control is relaying information to the storm spotters on what the radar is showing and where safe areas would be to set up and observe. During the net a mobile spotter disregards net controls instructions and takes on what can only be described as a "storm chaser" approach with the severe weather. He states "I know what I'm doing, don't tell me what to do!" and follows the storm on its path into County Y. He has become a "loose cannon." His behavior is dangerous to himself and makes the organization as a whole look bad. The next day, after the storm had passed, he admitted his wrong doing and the problems it caused. What fed this kind of behavior? Simply put, ego.

When working with volunteers stress is just one challenge that we face. We are also challenged by egos. All of us have an ego, some big, some not so big. During a stressful situation, such as severe weather response, egos can come to the surface. There may be control issues, "loose cannon" personalities, tempers, and all sorts of ugly behavior. These can create some embarrassing and at times dangerous situations. As coordinator or EC you must be ready and able to deal with these situations. During the severe weather event it is best not to confront ego issues head on. Address the specifics after the event is over. A skillful coordinator or EC will diplomatically and tactfully rein egos back in. It is better to rein in the loose cannon and keep them as a productive storm spotter than to anger them and encourage further behavior.

If you are in a position of managing any of your local storm spotters you have a challenging job. It is often thankless and carries with it a significant level of stress. Prepare yourself by training and learning how volunteer organizations work.

AFTER THE STORM

Once the storm has passed our role as amateur radio operators and SKYWARN storm spotters is not over: it enters a post-event phase. In the post-event phase we have four basic tasks. First is to continue our support of NWS efforts and contribute storm damage reports and any other reports that may have not been submitted during the event. Second is to submit appropriate reports to the EC, DEC, SEC and ARRL HQ on amateur radio activities in support of the NWS and severe weather response. Third is to conduct an after action report detailing our activities and response. Finally, fourth is to take part in any necessary debriefing that may occur. Let's look at each of these.

Damage Reports

At several points we've mentioned the importance of storm damage reports to the NWS. Following an event, meteorologists from the local WFO will set out to photograph and document damage related to the storm. Typically, though, local NWS offices do not have large staffs that can fully handle this task. This is where SKYWARN trained storm spotters can assist. Weather affects where we live. Storm spotters are familiar with their communities. We know what areas are accessible and inaccessible following a storm. And we can quickly spot damage from the storm. By utilizing the SKYWARN storm spotter network the local NWS office gains extra hands for the task of documenting how a severe weather event impacted our area. This information can be used in developing the local storm report, determining the nature of the event, determining the severity of the event and for future planning.

Post-storm reports are not limited to damage. Some spotters may not be able to relay in storm reports as events happen. Weather reports may be submitted after the event has ended. This information is also valuable. And some data such as total rainfall may not be fully known until after the event has ended.

ARRL Report Forms

In some areas the local EC may serve as the local amateur radio SKYWARN representative too. This may be in an informal, unofficial, capacity or it may be by appointment from the local NWS WFO. In other areas the amateur radio SKYWARN representative is a separate position from the EC. In either case, when amateur radio SKYWARN members are activated for severe weather it is a matter of concern for the local EC or, if there is no local EC, the DEC. For our purposes we are going to assume that the EC and amateur radio SKYWARN representative are two separate positions, held by two separate individuals. Severe weather response and SKYWARN activity can fall into two areas that amateur radio operators specialize in — public service and emergency communications. Because of this we will want to document our SKYWARN activity for our EC/DEC. Beyond the local level amateur radio severe weather response also concerns our section leadership, the SEC and SM and ultimately the ARRL.

The way we report our activities during severe weather are by using the standard ARRL Field Services Division (FSD) report forms. These are found on the ARRL's website on the amateur radio Public Service page, listed as "FSD forms."[3] There are several report forms that pertain to severe weather response. The specific forms we are going to cover in this section are:

• ARES Form 1— Public Service Activity Report

• ARES Form 2 — Monthly DEC/EC Report

• ARES Form 3 — EC Annual Report

• ARES Form 4 — Monthly Section Emergency Coordinator Report to ARRL Headquarters.

ARES Form 1 — Public Service Activity Report

So, let's start at the local level. The first report form we will cover is the Public Service Activity Report. We can consider SKYWARN activation as a public service activity because its very purpose is to provide information to the NWS in support of their mission to help safeguard life and property. This is an activity we want to be certain to document. We do this by using the ARRL ARES Form 1, and our report can be used in conjunction with others to show to Congress, the FCC, and public officials exactly how valuable a service amateur radio provides.

The ARES Form 1 allows us to document key data about our activation. First is the nature of our activity, there is a place to check for "Severe Weather/SKYWARN related." Next, we document places, people, and time of activity. The area for "Places or areas involved" should be specific and avoid using terms for locations that are only understood locally. Then indicate the number of radio amateurs involved (this may include Amateurs that are not part of the storm spotting group). When indicating date, keep in mind you may need to specify if these dates are local or UTC. Finally indicate the number of people hours. So, let's say you have three amateurs involved — K5DSG was active for two hours, KD5JHE three hours, and K5LMB one hour. Tthe total people hours is six hours.

The next step is to indicate spectrum used. This is important as it documents our use of our most valuable resource, the spectrum we have access to. Indicate all bands that were used for the activation. Then indicate the agency assisted; there is a place for National Weather Service.

Finally, we are going to provide information on the reporting party. There are no hard and fast rules on who should submit this report. Ideally it would be a designated, local amateur who was involved in the activation and the documentation of activities. That could be the local EC/DEC, net manager, or SKYWARN coordinator, but any amateur radio operator who has knowledge of the event can submit the report.

Once all of the documentation is complete, we will send the completed form to the local ARES EC. This form can be filled out online or can be printed out and mailed in. If the form is completed by someone other than the local EC/DEC, a copy should be sent to them as well. While there is no set time frame for the report to be submitted, it is best to do so as soon as possible after the event. The longer it is put off the more likely it will be that information will be omitted or less accurate.

The ARRL Emergency Coordinator

The ARRL Field Organization provides emergency communications management at local, district, and section levels with the Emergency Coordinator (EC), District Emergency Coordinator (DEC), and Section Emergency Coordinator (SEC) respectively. All of these positions fall under the general leadership of the Section Manager (SM). When it comes to amateur radio severe weather response, each of these has an interest.

The EC is appointed by the SM and works along with the DEC. The jurisdiction of an EC is typically a city or county. The EC "prepares for and engages in management of communication needs in disasters." While the EC is an ARES function, the position is part of severe weather response. This is largely due to the dual response role of ARES and SKYWARN. There are several specific duties of the EC that are listed on the ARRL website. We won't list them all here but will discuss duties that pertain to severe weather response.

The EC's duties include a variety of tasks. The EC is the point of contact for amateurs interested in getting involved in emergency communications. ECs help develop and provide training for amateurs. They develop and maintain working relationships between the local amateur radio community and the served agencies at the local levels. They provide input in the development of local emergency plans. They establish local emergency communications networks and regularly test them with exercises and drills. They serve as the emergency communications point of contact during a disaster. And they work to grow the ARES program. Basically, your local EC is your local amateur radio emergency communications manager. The EC can provide a great resource for local SKYWARN activity by assisting with training, exercises, and information, and being the point of contact between SKYWARN storm spotter activity and local officials.

There is one important duty of the EC that relates very directly to SKYWARN and storm spotter activity: "Report regularly to the SEC, as required." In the absence of an SEC the EC reports to the SM.

Emergency Coordinator Report Forms

There are two reports that the EC submits to the SEC/SM: the Monthly DEC/EC Report, ARES Form 2, and the EC Annual Report, ARES Form 3. On the monthly report, the EC reports the number of public service events and emergency operations conducted within his or her jurisdiction for the month. SKYWARN training classes that are sponsored by local amateur radio SKYWARN teams would constitute a public service event as long as the general public was invited to attend. SKYWARN activations would constitute an emergency operation, whether or not they are associated with an ARES activation. Besides the number of public service events and emergency operations the EC also reports the number of "person-hours."

On the annual report, the EC reports on the agencies that are served by the ARES group. Again, going back to the dual response nature of ARES and SKYWARN, the EC may want to include the NWS as a served agency through SKYWARN. If an area has someone other than the EC managing the amateur radio SKYWARN team, they should forward SKYWARN activation information to the EC to be included in monthly and yearly reports.

District Emergency Coordinator

Just like the function of the EC as a local emergency communications manager, the DEC serves in a similar capacity at an area level. The DEC's jurisdiction may be several counties or an area within a section. A key role for the DEC in regard to SKYWARN is that he or she serves as a local EC for counties or other jurisdictions that do not have one. If your area lacks an EC you may want to find out if you have a DEC to turn to for emergency communications guidance. As far as reporting goes, the DEC makes reports similar to those filed by the EC and submits them to the SEC.

Section Emergency Coordinator

The SEC is the amateur radio emergency communications coordinator at the section level. Like ECs and DECs, the SEC has a duty to submit reports on emergency communications activity within the section. The form that is used, ARES Form 4, is similar to the one used by ECs and DECs. It includes information on public service events and emergency operations conducted each month. The SEC reports this data directly to the ARRL.

AFTER ACTION REPORTS

"Lessons learned" has been a buzzword in the public safety community for several years now. Lessons learned from superstorm Sandy and the 2017 Atlantic hurricane season are still being discussed and as we experience other critical events, new lessons learned are added. Whether we respond to severe weather as SKYWARN spotters, ARES members, or as part of some other group, a valuable part of the activation process is assessing our performance, determining what went right or wrong and taking corrective action. Through this process we can identify objectives of the activation and determine to what degree they were met. The after action report (AAR) makes all this possible.

Writing AARs is standard in military, public safety, and emergency management environments. Any time there is a disaster, major public event, or critical incident an AAR will be written to assess the response. Disaster exercises and other training events also would warrant an AAR to assess response and determine if the exercise objectives were met. In this section we will go over what an After Action Report is, who writes and reads it, when it should be done, why it is done and how it is done. We will also look at the differences in After Action Reports written for critical incidents and exercises. As we go we will see how AARs can be used in amateur radio response to an emergency. And finally we will go over how to get your group started writing After Action Reports.

What is an After Action Report?

The US Army provides a great explanation on what an After Action Report is. In *A Leader's Guide to After Action Reports* they identify four key aspects to an After Action Report.[4] First, it is a discussion of an event. While the Army may be concerned with an event such as combat, for us an event could be a hazardous materials incident, earthquake or tornado. Second, it focuses on performance standards. Very seldom do amateur radio operators respond to an emergency situation without standards. In fact we pride ourselves and are often commended on our ability to excel in providing communications during a crisis. Third, an AAR explains to the reader what happened and why it happened. And finally the After Action Report provides insight to sustain strengths and improve on weaknesses. We are not perfect and mistakes will happen. We only fail at what we do when we do not learn from our mistakes and improve. By using a tool such as the AAR we can welcome mistakes in our response since it gives us a way of building from them. There may be differences on what an After Action Reports contains depending on who wrote it and what event it covered, but these basic goals of the report remain the same.

After Action Reports: Why Are They So Important?

The After Action Report (AAR) allows us to identify strengths and weaknesses in our response. When we identify weaknesses we can identify ways to improve our response in future emergencies. And over time AARs will make it possible to see trends and patterns in our response (both good and bad). AARs also allow those involved in the response to see the big picture of the event. They may see what other responders have done and discover solutions to problems or provide solutions to other responders.

There is another advantage to writing AARs. They can provide resources for responders outside of the affected area. By making them available to other agencies and groups it is possible to expand the knowledge base on response to emergencies. So going back to our ice storm example, when the final AAR is compiled it can be made available to ARES/SKYWARN groups and served agencies, such as the National Weather Service or Red Cross, outside the disaster area.

Photos of storm damage in After Action Reports can help the National Weather Service understand the extent of damage and what caused it. Tornado damage in Hoisington, Kansas, April 21, 2001. [Robert Haneke, WGØQ, photo]

Who Writes the AAR and Who Reads it?

The AAR may be a collaborative effort. Early on in any critical incident someone should be designated as the person responsible for compiling the After Action Report. This person should be familiar with planning the response, the organizations involved and the objectives of the response. Others may contribute to the report. Individuals involved in responding to the emergency can contribute by submitting reports on what happened, how they responded, problems encountered and objectives met. In the end these account may be compiled and put into the final report either in their entirety or incorporated into a summary of reports received. But for these reports to be accurate there must be a process in place to document events as they occur. Some ways of doing this are logs and journals, written messages, action plans for specific events, and public information and media reports.

This process can be used in writing an amateur radio After Action Report. Let's use a major ice storm as an example. Such an event can cover a wide area, affecting several states. In this case the final AAR may be a collaborative effort between the Section Emergency Coordinators in the affected area, each one putting together an AAR for their area with the final AAR being a compilation of each section's AAR. Of course to do this requires collecting data from the local responders. This can come from reports written by local Emergency Coordinators, District Emergency Coordinators, SKYWARN storm spotters, SATERN members and any other groups that may have responded. And just like any other event that may warrant an AAR there must be some form of documenting events as they occur. Amateurs are familiar with logs and can use these to keep track of reported events, messages and net activity. Field responders may keep a notebook journal to keep track of events and activities. And software is available to help keep track of resources and times. Needless to say, this can be a massive task. But the rewards of a good AAR are worth this critical follow-up work after the event. Later we will look at ways of streamlining the process of collecting this data.

So once it does get written, who reads it? An AAR is designed to be a useful tool to those who plan response, respond, and assist in future planning. The AAR would definitely be of use to all parties involved in the emergency response. This could be the individual storm spotters, net control operators, local and section leaders, and any number of served agencies that may have been involved. We can't look at this as a report isolated to amateur radio, but as a report of one component of a larger response. Sharing these findings with other responders can benefit the amateur radio response as well as their response.

What an AAR is not is a press release or any form of public information statement. It may contain information that isn't really suited for public release. This is not necessarily because there is anything to hide, but scrutiny by readers not familiar with all aspects of the emergency or response may not be entirely beneficial. If a public information statement is needed that is a separate matter handled by a public information officer or spokesperson.

When Should the AAR be Written?

Ideally the AAR should be written as soon after the event as possible. Depending on the size of the event this may be difficult to do. Documentation from the event needs to be gathered, responders or leaders may need to be interviewed, surveys may need to be conducted and perhaps workshops made up of responders may need to be held. A lot of work goes into gathering the information needed for a good AAR and it must be done in a timely manner. By waiting too long after the event we may forget some details, neglect documentation, or not provide enough detail in the report.

An amateur radio AAR is no different. Data must still be collected as soon as possible after the event, analyzed, and put into report form. During Hurricane Gustav an EOC staff member told me that after every hurricane they talk about getting everyone together to discuss the event, problems encountered and write a formal AAR. But each time, they told me, everyone is so worn out from the demands of the emergency that little is done and a far less formal AAR is eventually compiled. Events such as hurricanes and ice storms can take huge tolls on those who responded. But the event is not over when the threat is gone. The event is over when we have included in our recovery effort a plan for better responding to the next emergency.

What Goes in the AAR?

AARs come in different types. AARs for large scale events such as hurricanes and earthquakes can contain massive amounts of information, while events such as an isolated tornado or localized flood may contain less. And the goals of an AAR for a disaster exercise may be different than for a real emergency, since the purpose of the exercise is training. Regardless of the size of the event or whether the AAR is formal or informal the AAR contains certain key information.

The AAR begins with introductory information. This may include the type and location of the event. It should contain maps of affected areas (if available), a timeline of events, date/time of any proclamations or declarations (such as state of emergency, federal disaster area, and so on) and duration of event. Also included would be a general description of the event.

The second part covers a discussion of response. This discussion will look at different levels in the response effort. It will likely cover field level response, local government response, interactions within the operational area and the regional area, interaction with state level agencies and interaction with federal agencies. It will focus on planning, logistics, finance/

Cover sheet of the extensive After Action Report written following the "Hurricane Joshua" exercise in 2007.

Sample After Action Reports

After Action Reports (AARs) may be written following any severe weather event where local SKYWARN members are activated. AARs may also be written following exercises. The length and amount of detail of the AAR will depend a lot on the type of event or exercise being covered. A large scale, multi-jurisdiction event such as Hurricane Katrina or the 9/11 attack may require an AAR that is highly detailed, very lengthy, and requires coordinated effort from representatives of all the groups involved. A small scale, localized event may require far less. In either case certain basic information should be included.

Let's take a look at two sample AARs. The first (below), is written for a localized event that required a relatively small response. The second, in the Appendix at the back of the book, is for a communications exercise that involved several groups. You will find that they both contain similar basic information.

To help facilitate a good AAR each participant should be encouraged to write up a brief account of their experience during the event or exercise. The reports can help the writer of the AAR get a better overall picture of the response. One way to do this is to make a simple AAR form, whether paper or online, and ask participants to take a few minutes to fill it out as soon as possible after the event/exercise. For an example of the online AAR report, refer to the Lansing Area/Ingham County Amateur Radio Public Service Corps website.[5]

Lafayette County
Amateur Radio Emergency Service
After Action Report

Date of activity: April 27, 2011

Description of activity: SKYWARN activation for severe thunderstorm/possible tornado

Duration of activity: 1200-1530 CT

Amateur radio groups participating: Lafayette County ARES/SKYWARN

Served agencies participating: Lafayette County Emergency Management, National Weather Service

Describe served agency participation: Emergency management assisted in gather severe weather reports. All reports of severe weather were forwarded to the National Weather Service. National Weather Service issued weather watches, warnings, and statements.

Number of Amateurs participating: 8

List of amateurs participating: K5DSG, KE5NQP, K5LMB, W5MPC, KD5JHE, N5RB, KE5TMY, WB5VYH

Person-hours of Amateur service: 37

Describe goals of activity, both for served agency and serving group: Local SKYWARN members gathered severe weather reports and submitted them to local net control. Net control then relayed these reports in a timely manner to the National Weather Service via the amateur radio station at the National Weather Service Office. Requests to gather specific information were sent from the National Weather Service office via amateur radio nets to SKYWARN storm spotters in the local area.

Two reports indicated possible funnel cloud, one report of tornado, one report of flash flooding, three reports of hail greater than half an inch in diameter.

Did the event fulfill the goals? Yes.

What went well? Call up procedures went well; the severe weather net was run effectively. Easy communication with emergency management and law enforcement.

Areas needing improvement: Additional SKYWARN training is needed, one operator not involved with SKYWARN activities interrupted the net on several occasions, repeater coverage in some areas is not optimal.

Lessons learned: Study repeater coverage area and look at ways to improve, add a relay station to assist net control during times where traffic volume is high, coordinate efforts with neighboring county ARES/SKYWARN.

General comments: Local SKYWARN members did a great job with severe weather reports.

Ideas for future exercises: A severe weather exercise coordinated with local emergency management, district ARES resources, University of Mississippi and the National Weather Service would be useful in the near future.

END OF REPORT

---

administration and multi/inter-agency coordination (this should be familiar from ICS training). Granted, this may be beyond the scope of an amateur radio response. But this is a section that can help us focus on our response as part of a coordinated effort at local, section, division and regional levels. There are times, though, that this may not be needed, such as when the event is isolated and does not involve participants at anything higher than a local or state level.

Next the AAR covers the participants and systems involved in response. This is where we cover the individual agencies and groups involved. We may first start off by addressing mutual aid systems that were involved. During a hurricane in the Gulf Coast area a memorandum of understanding may be utilized to help coordinate efforts between affected sections in the Delta Division. This section will also address the participants in the response. Most times amateur radio is one part of a large-scale response. Other players may include public utilities, Red Cross, National Weather Service, Salvation Army, the media, and public safety. There may also be interaction with other levels of response. Assessment of these interactions is important. When amateur radio responds we do not go alone. We generally assist other agencies and groups and an assessment of our interaction with them is critical. And, as you can probably guess, they are assessing their interaction with us.

Next the AAR will address training needs. This is where we look for areas in our response that can be improved on by further training. There is no such thing as a perfect response. If we get into the mindset that our actions during a disaster have no room for improvement then we are setting ourselves up for an even bigger disaster in the future. Talk with all involved in the response and get feedback on what went wrong as well as what went right. Problems may be small or large but either way should be addressed so they don't come back to haunt us in the future. This is also a good spot to focus on some training that the group may need. Does your group have problems with radioing in weather information that is not reportable? Identify that issue here and discuss ways that the group can be trained on appropriate reports.

What Do We Do With the AAR Once It is Written?

After the AAR has been properly distributed to those that need to read it the AAR should be kept on file securely; cloud storage is highly recommended. Remember this is a tool for future planning: it must be accessible for future use. One way we can keep an AAR valuable is by conducting a future review. Let's say the AAR was written by the local Emergency Coordinator for a tornado event. Naturally all SKYWARN and ARES members will read it as soon as it is available. And it is likely it may be looked at by the local emergency management director. If it goes on the shelf and is forgotten about at this point then it hasn't served its true purpose. Now let's say a couple months later the SKYWARN and ARES groups are going to conduct an exercise to test their response to severe weather. The EC re-reads the AAR and applies it to how the test is going to be conducted. This helps give some guidance to developing an exercise that encourages learning from real life experience. Now let's say the same area is facing the threat of a tornado a year later. The EC and emergency manager can look over the AAR and find reminders of what worked, what didn't work and eventually see if additional training and exercises paid off.

The AAR is a concept well rooted in emergency management, military, and public safety, but it is not necessarily exclusive to these fields. amateur radio can make valuable use if AARs as a tool for severe weather response.

Debriefing

A debriefing is another useful tool for assessing our response to severe weather. The debriefing typically involves more than just the amateur radio storm spotter group. It will likely involve local emergency management and/or public safety officials. It is also possible a debriefing may be done with representatives from the local NWS office.

The purpose of the debriefing is to review the effectiveness of the response and address issues of concern. Being present at the debriefing allows you to directly answer questions that may come up about the amateur radio response. It also gives you a forum to bring up issues of concerns for others involved and to ask questions. You will want to bring with you to the debriefing information from the event. Do not rely on memory and log books alone. Throughout the event keep a separate diary of issues for the debriefing session, a "Debriefing Diary." This should contain issues not appropriate for the station log book, information you will need to retain if the log book has to be handed over to someone else, and information about specific events, times, places, and other information that needs to be mentioned. Here are some other items for the Debriefing Diary:

- What was accomplished?
- Is anything else still pending? Note unfinished items for follow up.
- What worked well? Keep track of things that worked in your favor.
- What needed improvement?
- Ideas to solve known problems in the future.
- Key events.
- Conflicts and resolutions.

Remember to focus on constructive criticism during the debriefing and not attacks on actions taken, finger pointing, or casting personal blame.

REFERENCES

[1]Federal Emergency Management Agency, "Emergency Planning," Independent Study IS-235, **training.fema.gov/is/**.

[2]National Climatic Data Center (NCDC): **www.ncdc.noaa.gov**

[3]ARRL Amateur Radio Public Service page: **www.arrl.org/public-service**

[4]United States Army, "A Leader's Guide to After Action Reviews," September 1993.

[5]Lansing Area/Ingham County Amateur Radio Public Service Corps website: **www.lansingarpsc.com**

Appendix 1

Weather Books for the Storm Spotter

GENERAL METEOROLOGY/WEATHER

Glossary of Weather and Climate by Ira Greer

Peterson First Guide to Clouds and Weather by V. Schaefer, P. Pasachoff, R. Peterson

The AMS Weather Book the Ultimate Guide to America's Weather by Jack Williams

Weather by Storm Dunlap

The Book of Clouds by John Day

The National Audubon Society Field Guide to North American Weather by David Ludlum

The Weather Wizard's Cloud Book: A Unique Way to Predict the Weather Accurately and Easily by Reading the Clouds by Louis Rubin, J. Duncan and Hiram Herbert

Weather: Air Masses, Clouds, Rainfall, Storms, Weather Maps, Climate by Paul Lehr and Will Burnett

Weather Basics by Joseph Basalma and Peter R. Chaston

WEATHER ANALYSIS AND FORECASTING

Basic Essentials Weather Forecasting by Michael Hodgson

Storm Chasing Handbook by Tim Vasquez

Weather Analysis by Dusan Djuric

Weather Forecasting Handbook by Tim Vasquez

Weather Forecasting: Rules, Techniques and Procedures by George Elliot

Weather Map Handbook by Tim Vasquez

Weather Maps: How to Read and Interpret All the Basic Charts, 3rd edition, by Peter R. Chaston

SPECIFIC STORMS

Category 5: The 1935 Labor Day Hurricane by Thomas Neil

Hurricanes! by Peter R. Chaston

Roar of the Heavens by Stefan Bechtel

Storm Warning: The Story of a Killer Tornado by Nancy Mathis

The Tri-State Tornado: The Story of America's Greatest Tornado Disaster by Peter Felknor

Thunderstorms, Tornadoes, and Hail by Peter R. Chaston

TEXTBOOKS

Basic

A World of Weather Fundamentals of Meteorology by Jon Nese and Lee Grenci

Meteorology Today: An Introduction to Weather, Climate, and Environment by C.D. Ahrens

Advanced

Mesoscale Meteorology and Forecasting by Peter Ray

Meteorology for Scientists and Engineers by Roland Stull

Mid-Level Weather Systems by Toby Carlson

Appendix 2
Weather Websites

National Weather Service
www.weather.gov

National Hurricane Center/Central Pacific Hurricane Center
www.nhc.noaa.gov

National Severe Storms Laboratory
www.nssl.noaa.gov

NOAA/NWS Storm Prediction Center
www.spc.noaa.gov

NOAA Weather Radio
www.weather.gov/nwr

NWS Aware Report
www.weather.gov/publications/aware

NWS JetStream
www.weather.gov/jetstream

NWS SKYWARN
www.weather.gov/SKYWARN

Canadian Hurricane Center
weather.gc.ca/hurricane/index_e.html

Commercial Sites
AccuWeather.com
www.accuweather.com

Allison House
www.allisonhouse.com

Google Earth
www.google.com/earth/

GRLevelX
www.grlevelx.com

RadarScope
www.radarscope.app

Spotter Network
www.spotternetwork.org

Stormpulse
www.stormpulse.com

The Weather Channel
www.weather.com

WeatherBug
www.weatherbug.com

WeatherTap
www.weathertap.com

Weather Underground
www.wunderground.com

Organizations
American Meteorological Society
www.ametsoc.org

National Hurricane Conference
www.hurricanemeeting.com

National Weather Association
www.nwas.org

Education
MetEd
www.meted.ucar.edu/index.php

Weather World 2010 (WW2010)
ww2010.atmos.uiuc.edu/(Gh)/home.rxml

Amateur Radio and CWOP
ARRL
www.arrl.org

APRS
www.aprs.org, aprs.fi

Citizen Weather Observer Program (CWOP)
www.wxqa.com

D-STAR
www.dstarusers.org

DMR
dmr-marc.net

EchoLink
www.echolink.org

Hurricane Watch Net
www.hwn.org

SATERN
qso.com/satern/

VOIP Hurricane Net/Weather Net
voipwx.net

WX4NHC
w4ehw.fiu.edu

Yaesu System Fusion
systemfusion.yaesu.com/

Appendix 3
A Local SKYWARN Operations Manual

At the local level a SKYWARN group may develop a local operations manual. Doing so will add some structure and strength to the organization. It also helps local SKYWARN members to know what to do, who to report to, and how local net operations are conducted. Alachua County (Florida) SKYWARN developed such a manual in June 2000. While some simple basics were borrowed from a very early version of the *ARRL Emergency Coordinator's Handbook* and elsewhere, the manual was designed largely using NASA's Mission Control Center in Houston, Texas, as a model. This idea allows for tasks to be delegated or assigned and permits multiple things to be done at the same time simply by giving specific jobs to specific positions. Basically, many of the tasks that an NCS might be required to take on could be assigned to others so they can be accomplished faster and much more efficiently. Delegating tasks to others relieves pressure on the NCS and helps to alleviate chances for confusion, frustration, and lowered morale from taking on a task that can get very tedious when things start to get fast-paced, and increases speed of situation handling.

Developing a local operations manual can be one of the biggest challenges a local SKYWARN group can face. Severe weather isn't the same everywhere and neither is the response. For example, some areas may have civilian organizations and others may have started within local emergency management. Others may even be a part of a military structure — say, operating under a MARS-related organization. (Some MARS organizations have also used the Alachua County SKYWARN SOP manual as a template.) For this reason, it is impossible to simply have one operations manual that will work for all locations. The manual developed in Alachua County, however, does give us a template to work with in developing a local manual.

The manual is divided into two sections: content and appendices. Content covers the basic information about the organization, membership, training, activation, net control, and safety. The appendices are a reference section for storm spotters. Let's go over what a model operations manual will cover.

CONTENT

Purpose — This is a statement that addresses the purpose of SKYWARN and the local SKYWARN group. It also covers the purpose of the manual.

Limitations — This sets out limits within the organization. It may spell out limits placed on the manual, members, and purposes of the organization. This is also where we would clearly state limitations on mobile spotters to not act as storm chasers.

Organizational Structure — Local SKYWARN organizational structure should be clearly spelled out. Positions included may include: SKYWARN Coordinator, Assistant SKYWARN Coordinator, training managers, net managers, public information, recruiting, and technical managers. The number of positions will depend upon the size of the SKYWARN group.

Membership — Indicate membership requirements such as Amateur Radio license, NWS SKYWARN training, membership cards, and so on.

Training — Required and expected training should be covered here. Don't forget to look beyond the SKYWARN class, cover training nets, local training meetings, SETs, and any other training functions.

Alert Authorization — This should clearly indicate how the local SKYWARN group is activated and who has the authority to activate the group. Remember that there are times that severe weather will develop with little to no warning. Be sure to address activation procedures from local emergency management or SKYWARN coordinator. Also include information on the repeater to be used for SKYWARN activation.

Mobilization — This covers how members will be advised of SKYWARN activation. Include all methods that may be used, such as radio, telephone, cell phone, pager, or e-mai. Also address what SKYWARN members are to do when mobilized, typically this will be to check into the SKYWARN net and await further instructions. Another topic within mobilization will be levels of activation such as stand by alert or emergency alert. Your group may use other indicators such as conditions green, yellow, and red.

Operation Procedures — This section explains how operations are to be conducted during SKYWARN activation. Include specific operation instructions for each level of alert, reporting procedures, how and when the net will be identified, and how and when the net will be secured.

Net Procedures — The SKYWARN net can operate like a well-oiled machine or can be total chaos. How you net goes depends on how well it is structured and the instructions provided for net control stations and stations checking in. A portion of your manual should include all instructions for conducting a directed severe weather net. There are two areas that should be addressed. First address net control stations. Second address the storm spotters that will be checking in.

Safety — Safety must be taken seriously at all times. The local SKYWARN group should have a designated Spotter Safety Officer. The manual should spell out the SSO's duties which typically include: monitor radar activity, monitor spotter reports to identify safety issues in the area, plot these events on a map, and when necessary advise net control of dangerous situations for spotters and provide information on the safest and fastest route away from the danger area. Additionally the SSO will monitor for unsafe activity from spotters. This information should be provided to the SKYWARN Coordinator or other designated individual for follow up action.

Frequencies — The manual should list the specific severe weather net frequencies that will be used by the local SKYWARN group. List only frequencies used for local SKYWARN communication; primary repeater or simplex frequency and any back up frequency should be listed. Frequencies for NOAA Weather Radio, area repeaters, public safety, and other services should be listed in an appendix.

Definitions and Miscellaneous — List here any acronyms or terms that are frequently used by the group and their meaning. Also provide definitions for NWS terms such as watches, warnings, and advisories. Reportable criteria may also be listed here. This may also be a place where you can list "don't do" items.

APPENDICES

The appendices are references for the SKYWARN manual. Specific contents of the appendices will vary but here are a few ideas that you may want to include.

Related Government Contacts — Include local, state, and NWS contacts. Also include served agencies such as the Red Cross and Salvation Army.

Storm Spotter Roster — Include names, call signs, and contact information.

Net Preambles — A script for opening the net, net continuity, announcements, roll call, and securing the net.

Phonetic Alphabet

Frequencies — local SKYWARN frequencies, NOAA Weather Radio frequencies, any simplex frequencies used.

Spotter's Checklist — Items that a spotter may need, especially for mobile spotting.

Spotter Locations — List known good spotting locations. Be sure to emphasize spotter safety.

Cities in the Area — List area towns and cities and their respective counties.

Maps — Provide any appropriate maps: city, county, district, or other applicable areas. Don't forget to include your local NWS County Warning Area.

Reporting Form — A form listing specific reportable weather events.

Net Control Log Form — This log sheet is used by net control to keep track of net activity. It should include: Called in time, call sign, name, time in, time out, location, and comments.

Charts — Beaufort Scale, Fujita Scale, Saffir-Simpson Scale, hail size chart.

Recommended Books — Books for the storm spotter

SKYWARN Products — Caps, jackets, badges, magnetic signs, and so on.

Revision History — Make sure to indicate any revisions to the manual; what was revised, when, and any other pertinent information.

The following pages show an actual SKYWARN Operations Manual prepared by the National Weather Service Forecast Office in Albany, New York.

SKYWARN OPERATIONS MANUAL

National Weather Service Forecast Office
Albany, New York

For Eastern New York – Southern Vermont – Berkshire County Massachusetts and Litchfield County Connecticut

Fourth Edition – September 2012

I. INTRODUCTION and ORGANIZATION

1.1 Purpose of This Manual

This manual is designed to be used as a reference guide for SKYWARN operations in the jurisdictions under the National Weather Service (NWS) Albany, NY, Forecast Office's warning area of responsibility. As a reference, it will never be complete and it is expected that individual counties will supplement this manual with their own internal policies and procedures, keeping the bi-directional flow of critical information moving smoothly and un-impeded.

The Albany area SKYWARN has a large area of responsibility which roughly extends from the north end of Lake George to near Newburgh and from just east of Utica to the Vermont-New Hampshire border and then south through Berkshire County, MA and Litchfield County, CT.

Training cannot, and should not, take place "on the job" during severe weather. Proper training is essential for the effective flow of information between SKYWARN spotters and the NWS and/or emergency management personnel. This includes training for spotters as well as net control volunteers. To be effective, everyone in the SKYWARN "system" needs to know their roles BEFORE severe weather strikes.

1.2 Purpose of SKYWARN

SKYWARN is the NWS national program of trained volunteer severe weather spotters. SKYWARN volunteers support their local community and government by providing the NWS with timely and accurate severe weather and flood reports. These reports, when integrated with modern NWS technology, are used to inform communities of the proper actions to take as severe weather threatens. SKYWARN, formed in the early 1970's, has historically provided critical severe weather information to the NWS in time to get the appropriate warnings issued. Thus the key focus of the SKYWARN program is to save lives and property through the use of the observations and reports of trained volunteers.

Each NWS forecast office runs its own SKYWARN program. It is a goal and a challenge to continually improve the SKYWARN system and to integrate new technologies and procedures to best fulfill SKYWARN's mission of saving lives and property. This includes but is not limited to; linked repeater systems, IRLP and VOIP (Voice Over Internet Protocol).

1.3 Role of Amateur Radio in SKYWARN

Amateur radio has been, and always will be, a critical component of the Albany area SKYWARN program. In the eastern New York and western New England area we are extremely fortunate to have a large number of trained SKYWARN spotters who are also amateur radio operators. This dual role for amateur radio operators is a natural result of their inherent interest and fascination with natural and scientific phenomena (especially the weather!) and with cutting edge technology such as Doppler radar and lightning detection devices. When this fascination is combined with the ability and desire to be trained to communicate severe weather observations via amateur radio in a professional and effective manner, the synergy is hard to duplicate. Finally, amateur radio operators have a long history of using their training, skills and equipment in uncompensated public service to help the community at large, which is precisely the focus of the SKYWARN system.

The close working relationship between the NWS and the amateur radio community provides many special benefits to each group. These benefits are highlighted in the following goals for the SKYWARN Amateur Radio operations:

1. To provide the NWS with timely and accurate severe weather reports via amateur radio. This includes both; incoming reports of severe weather per the NWS criteria; and amateur radio operators making observations at specific locations in response to a NWS request. For example, amateurs have often been asked to monitor river and creek flooding situations at certain critical points.

2. To create and maintain an organized communication network for passing critical severe weather traffic in a timely fashion to and from the NWS in the event that normal communications have been interrupted. The NWS has lost normal communications services in the past and it is likely that the SKYWARN Amateur Radio Net would be activated in future communications emergencies.

3. To disseminate warnings and weather statements issued by the NWS to the amateur radio community. Every attempt is made to read special and severe weather statements issued by the NWS over the SKYWARN Net, as well as updated storm movement information to keep amateurs informed of developing situations and to practice for situations when normal communications channels fail.

4. To organize and train amateur radio operators to prepare themselves and their families for disaster or emergency weather related situations so that they may be available to assist in emergency net operations. This preparedness training is critical if the SKYWARN system is to be expected to operate reliably during true emergency situations.

1.4 Organizational Structure of SKYWARN

SKYWARN is NOT a club. It is a true volunteer public service whose membership is open to all who wish to participate. All reports of severe weather through the SKYWARN system are appreciated. Scripts have been set up to outline the NWS criteria for severe weather on which observations are requested so that untrained observers may participate. Despite the scripts, all net participants are strongly encouraged to take advantage of the excellent, interesting and free training provided by the NWS covering basic and advanced SKYWARN training as well as organized specialized courses on winter storms and floods.

The structure of SKYWARN under the Albany NWS jurisdiction is as follows:

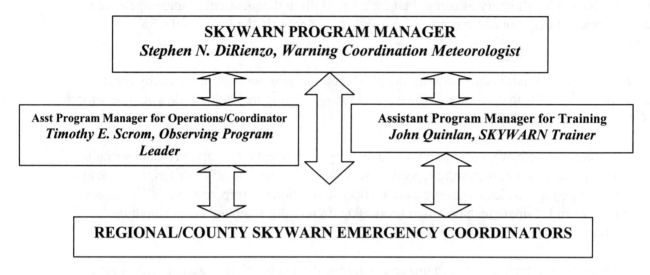

1.5 SKYWARN's Relationship to ARRL/ARES/RACES/REACT

The Amateur Radio operator's participation in the SKYWARN program is formally acknowledged and encouraged in a Memorandum of Understanding (MOU) between The American Radio Relay League (ARRL) and the NWS. This agreement indicates that the ARRL will encourage its local volunteer groups operating as the Amateur Radio Emergency Services (ARES) to provide the NWS with spotters and communicators as requested by the NWS during times of severe weather.

Many civil disasters are a direct result of severe weather and/or are exacerbated by severe weather. Accordingly, the NWS may utilize the SKYWARN amateur radio operators not only to obtain and disseminate severe weather observations and warnings, but may also use the amateur radio operators to maintain close coordination with Emergency Managers under Amateur Radio Emergency Service (ARES) and Radio Amateur Civil Emergency Service (RACES).

Radio Emergency Associated Communications Teams (REACT) also support SKYWARN. REACT nets may take reports of severe weather and relay them to the NWS either by normal communications modes (phone, FAX etc.) or by linking up with a REACT member who is also an amateur radio operator who can relay the severe weather information to SKYWARN Net Control through the SKYWARN amateur radio frequencies. Although it may take some creativity and organization, the goal is to include all groups in the SKYWARN system who wish to participate.

1.6 Role of SKYWARN Net Control

SKYWARN Net Control is the critical role in any SKYWARN activation. It is a role that will always challenge all of an amateur radio operator's communications and technical skills. It is also an extremely responsible role in that the safety of lives and property may rest on the amateur's skills. Although this role is challenging, with proper training and experience, this role can also be extremely rewarding when a job is successfully completed.

It is the purpose of this manual to provide the general guidelines for SKYWARN operations. While consistency of procedures from net to net is important, no two SKYWARN activations will be exactly the same. Therefore, the net control operator has the authority and responsibility to do everything necessary, within FCC rules, to ensure that the SKYWARN mission is performed to the best of his or her abilities.

It is essential that SKYWARN net control operators be familiar with the operation of the SKYWARN Amateur Radio Station, as well as NWS procedures, to be able to do an effective job.

1.7 Role of the SKYWARN County Coordinator

The SKYWARN County Coordinator organizes the operation of the entire SKYWARN Amateur Radio community within their county to ensure operation in accordance with the goals of the NWS. Specific duties of the SKYWARN County Coordinator include, but are not limited to:

1. Keeping a set of recommended protocols and an operating manual up to date.

2. Coordinating simple, effective and efficient procedures for passing traffic between local SKYWARN nets and the NWS forecasters.

3. Sharing information, ideas, and protocols with other SKYWARN programs throughout the country to develop the best possible local SKYWARN program.

4. Coordinating the activities of SKYWARN with ARES, RACES, REACT, and government agencies to best fulfill SKYWARN's goals.

This volunteer position is appointed by the NWS SKYWARN Program Manager to ensure that the person chosen can work well with the NWS forecasters and management as well as the amateur radio community. The person chosen for this position must possess superior coordination and communication skills and should be readily available to the NWS.

1.8 Role of the SKYWARN Regional Coordinator

The SKYWARN Regional Coordinator organizes the operation of the entire SKYWARN Amateur Radio community within those counties for which the Albany NWS office is responsible. Specific duties of the SKYWARN Regional Coordinator include, but are not limited to:

1. Keeping a set of recommended protocols and an operating manual up to date to ensure compliance.

2. Coordinating simple, effective and efficient procedures for passing traffic between local SKYWARN nets and the NWS forecasters.

3. Sharing information, ideas, and protocols to develop the best possible local SKYWARN program.

4. Coordinating the activities of SKYWARN with ARES, RACES, REACT, and government agencies to best fulfill SKYWARN's goals.

5. Assist County EC's in the execution of their duties and act as a liaison with the NWS and other agencies.

This volunteer position is appointed by the NWS SKYWARN Program Manager and/or ARES, to ensure that the person chosen can work well with the NWS forecasters and management as well as the amateur radio community. The person chosen for this position must possess superior coordination and communication skills and should be readily available to the NWS.

The SKYWARN Regional Coordinator is also responsible for making sure that at least one "coordinator" is on duty at all times to receive the NWS notification and to take appropriate action as requested by the NWS. This will often involve passing the SKYWARN activation instructions and trained net control volunteer lists from coordinator to coordinator when an out of town trip is expected. It is imperative that this position be covered at ALL TIMES!

II. ACTIVATION of SKYWARN

2.1 NWS Decision to Activate SKYWARN

The NWS Albany Forecast Office activates SKYWARN when severe weather is expected to affect its area of warning responsibility. See map in Appendix D of this manual. SKYWARN is activated for many forms of anticipated severe weather including tornadoes, severe thunderstorms, hurricanes, floods, and major winter storms.

2.2 Activation Time Frames and Requested Staffing

For short lead time events (i.e., severe thunderstorms, tornadoes. and flooding), SKYWARN is activated when the WATCH is issued or when severe weather is probable. The lead time may vary from zero (0) to six (6) hours. Volunteers are usually requested to staff the NWS SKYWARN Amateur Radio station when a Severe Thunderstorm Watch or Tornado Watch has been initiated for our County Warning Area (CWA). The minimum number of counties in the Watch should be at least eight to call in operators. Contact the Net Control Call Up Tree members (called in order) to arrange for a severe weather operator. SKYWARN operations could last for up to twelve hours for short term events.

During long lead time severe weather events such as hurricanes, stream and river flooding, and winter storms, SKYWARN is activated when the WARNING is issued. Lead time may be anywhere from zero (0) to twelve (12) hours. Requests to staff the SKYWARN Amateur Radio Station will depend on the forecaster's assessment of the nature of the storm and the storms potential. Long-term events may cause SKYWARN to be activated for extended periods of time, possibly measured in days, such as during the Blizzard of '93.

2.3 NWS SKYWARN Activation Steps

Once NWS forecasters have made the decision to activate SKYWARN, the following action is taken by the forecasters:

The Hazardous Weather Outlook message is updated with the specifics and the last segment of the message, "Spotter Information Statement" is changed accordingly with activation information naming specific counties to be activated. In addition notify the Regional SKYWARN Coordinator or designate when a Severe Thunderstorm or Tornado Watch is issued even if it does not include enough counties to necessitate activating Net Control in the office.

The Hazardous Weather Outlook message is a permanent part of the 24 NOAA Weather Radio broadcast cycle and is also on our web page. Generally, if there is a threat of severe weather, it will be contained in the first segment of the message and the Spotter Information Statement will state that SKYWARN activation may be necessary later in the day. It is issued routinely by around 5 AM daily and is updated as necessary.

This message alerts SKYWARN spotters and emergency managers to be on the lookout for severe weather and to be ready to pass reports to NWS by phone if nets are not operating.

Most of the watches and warnings that cause SKYWARN to be activated are tone-alerted and will activate weather alert radios. SKYWARN participants are encouraged to obtain radios with this feature. The tone alert feature is tested each Wednesday between 11 AM and Noon by the NWS. Please make sure that your tone alert is functioning properly!

Below is a listing of the NOAA Weather Radio transmitters that we operate from our office in Albany NY. Also included is the general listening area as well as the Counties for which we issue tone alerts on that particular transmitter. If you have a Specific Area Message Encoder capable receiver, this information is critical for your use in programming your receiver. The numerical codes necessary to enter into the receiver are listed on our web page under the NOAA Weather Radio section.

Transmitter	Station	Frequency	Listening Area	Counties
Gore Mountain, NY	KSC-43	162.450	upper Hudson Valley, southern Adirondacks & west central VT	Addison & Rutland VT. Essex, Fulton, Hamilton, Saratoga, Warren & Washington NY.
New Scotland / Albany NY	WXL-34	162.550	Capital District, eastern Mohawk & Schoharie Valleys	Albany, Columbia, Fulton, Greene, Montgomery, Rensselaer, Saratoga, Schenectady & Schoharie NY.
Poughkeepsie / Highland, NY	WXL-37	162.475	mid-Hudson Valley & Litchfield Hills	Litchfield CT. Columbia, Dutchess, Greene, Orange, Putnam, Sullivan & Ulster NY.
Fairfield / Middleville, NY	WXM-45	162.425	Southern Adirondacks, Mohawk & Schoharie Valleys	Fulton, Hamilton, Herkimer, Lewis, Madison, Montgomery, Onieda, Otsego & Schoharie NY.
Ames Hills / Marlboro VT	WXM-68	162.425	southern VT, western MA & southwest NH	Bennington, Windham & Windsor VT. Berkshire, Franklin & Hampshire MA. Sullivan & Cheshire NH.
Egremont, MA	WXM-80	162.450	Taconics, southern VT, western MA & Litchfield Hills	Bennington & Windham VT. Columbia & Rensselaer NY. Litchfield CT. Berkshire, Franklin, Hampshire & Hampden MA.
Mt. Greylock, MA	WWF-48	162.525	Taconics, southern VT, western MA & Litchfield Hills	Bennington & Windham VT. Columbia & Rensselaer NY. Litchfield CT. Berkshire, Franklin, Hampshire & Hampden MA.
Cornwall CT	WWH-33	162.500	northwest CT & adjacent areas of NY / MA	Fairfield, Hartford, Litchfield & New Haven CT. Dutchess & Putnam NY. Berkshire & Hampden MA.

2.4 SKYWARN County EC Activation Steps

1. The SKYWARN County EC, or his designate, receives notification and calls the NWS as

necessary.

2. The EC is briefed by the forecaster on:

    a. Nature of expected severe weather;

    b. Expected onset of severe weather (immediate or later in day);

    c. Expected duration of event;

3. The SKYWARN County EC or designate, assesses the situation and takes the appropriate action for his/her county SKYWARN operations, contacting personnel to run a net, and if 8 or more counties are within a Severe Thunderstorm or Tornado Watch the SKYWARN Regional Coordinator or designate contacts persons within commuting distance of the Albany NWS to operate the NWS SKYWARN station. The SKYWARN County EC or designate is to get information to NWS Albany by the quickest means possible.

4. Meanwhile, if a SKYWARN County EC or designate determines that a severe weather report, such as a funnel cloud, needs to be made known to the forecasters IMMEDIATELY, and the Amateur Radio volunteer has not arrived at the NWS, please make sure that the NWS is informed of the traffic by quickest means possible.

2.4.1 Net Control Operator Personnel List

The SKYWARN County EC's will maintain a list of personnel available at specific times to run a SKYWARN net in their county.

The SKYWARN Regional Coordinator will maintain a list of operators for the NWS SKYWARN station, and provide this list to the NWS. The operators would be chosen on an as available basis. The goal is to have numerous operators on the list, many of whom are likely to be available in the afternoons to cover SKYWARN activations for severe weather. The NWS must have the ability to contact operators at any time and therefore should have the amateur's home phone, work phone, car phone, FAX, and pager numbers.

III. SKYWARN PROCEDURES AND PROTOCOLS

3.1 Operating Rules for County or Regional SKYWARN Nets

The normal net protocols will be utilized on all nets. Above all else, common courtesy must be exercised. The NWS relies on the SKYWARN spotters for critical information which could impact life and property. An organized effort to channel this information to the NWS must be in place in order for all to benefit from the SKYWARN operation.

As mentioned elsewhere in this manual, the County EC or his/her designate, is responsible for the SKYWARN net operation. It is assumed that permission has been granted, prior to SKYWARN operations, by the various repeater trustees, licensees or owners, for the use of said repeaters. It is the responsibility of the County EC's to gain that permission. It is also the responsibility of the County EC's to ensure that backup communications are available, such as other repeaters or other bands.

Unless the controlling interests of specific repeaters feel it necessary, or unless the conditions warrant it, no repeater should be dominated by SKYWARN activities. In most instances, normal amateur communications can continue with only an informal SKYWARN net in operation. Only during particularly severe weather such as a tornado on the ground or significant damage from severe thunderstorms or flooding, should the SKYWARN net transition to a formal (directed) net.

The SKYWARN operator of the NWS SKYWARN station will, when roving for reports, check into specific nets for reports or to read warnings or statements, following net protocols. If no net is in progress, the operator will make a general call for information, or make an announcement that specific information is available and ask if anyone was present to receive it. The operator will not initiate any net. The SKYWARN State Coordinators shall follow the same procedures.

There will be instances when communications problems will prevent SKYWARN spotters from communicating with their own counties and or the NWS. In that case, it is highly desirable that those reports be handled in the same way as reports for the Albany NWS. In so doing, the report will be relayed by the Albany NWS to the NWS office that is responsible for the area the spotter was reporting from, i.e., Burlington, Boston, Binghamton, Buffalo or Brookhaven (NYC). This is a common practice throughout the NWS. In the event of a widespread severe weather outbreak, the SKYWARN Regional Coordinator or designate will specify a coordinated net to accept reports from local nets.

3.2 Local Weather Nets/Self Activation

The weather is very difficult to predict! Local severe weather, such as flooding or severe thunderstorms, may develop suddenly without the NWS issuing a watch or warning, or be too localized for the NWS to activate SKYWARN.

The following is the recommended procedure for implementing local area weather nets.

The activation of a local area weather net should be coordinated on the local level with an ARES

EC and the repeater licensee, preferably in advance of the weather emergency. To be successful and to serve the NWS in the best possible manner, the program needs to be self policing. Therefore, the structure should be similar to any SKYWARN net where there is one Net Control station and one assistant to make sure that severe weather reports are relayed to the forecasters.

Upon receiving reports of a serious local weather situation developing, the Net Control station should contact the NWS lead forecaster by telephone to:

1. Relay the weather information.

2. Confirm that the NWS has not activated SKYWARN and will not do so (AFTER having listened to available sources).

3. Receive a request from the lead forecaster that a localized area of severe weather is in your location and that reports are needed. Please give the lead forecaster your name, call sign and telephone number and indicate that you are the contact person for running a local weather net on a particular frequency in a particular area. The forecasters may wish to listen to the net "live."

Please designate the area as a local area weather net and not as a SKYWARN Net. This notifies participants that any critical weather information needs to be relayed to the NWS by telephone and not by amateur radio as, most likely, there is no one listening to the Amateur Radio Station at the NWS.

If SKYWARN is activated after a local area weather net is in progress, the local area net should transition to a SKYWARN Net.

To be effective, the participants in the local area weather net should have completed SKYWARN Basic Spotter Training.

3.3 Handling Non-Severe Weather Reports

Many of the reports received are for non-severe weather. Please be courteous to the report giver and note the amateur's location as you may need to contact the amateur(s) if the storm moves in their direction. As the moment dictates, it may be necessary in periods of extremely severe weather to only take reports from specific areas of interest to the NWS or only reports of severe weather. If the situation arises, please do not be shy about informing the net participants of the exact nature of the information needed and that the only report you will take must meet the severe weather criteria. Please indicate when the net is reopened for all traffic.

IV. NWS STATION OPERATOR PROCEDURES

4.1 Behavioral Rules

4.1.1 DO NOT Bring Children with You!

The NWS Forecast Office is NOT the place for children or sightseers during emergencies. The NWS operations area is not large and is filled with expensive and delicate equipment. Please DO NOT bring people to the NWS who will distract you or the NWS from doing the best possible job. The NWS would be pleased to give your family a tour of the NWS facility at a quiet weather time and upon prior arrangement.

4.1.2 NWS Forecast Office Operating Conditions

When SKYWARN is activated the NWS is usually operating in a high tension and critical weather mode due to weather conditions. This means:

1. Any distractions or interruptions of NWS or SKYWARN operations may mean the loss of life or property.

2. Sensitive information such as severe damage or loss of life may be openly discussed within the NWS office and should not be repeated by SKYWARN volunteers outside the NWS.

3. TV and/or news crews may be present in the forecast office or at a remote operating site. Please refer all media questions to the NWS Severe Weather Coordinator on duty.

4.1.3 No More Than Three (3) Volunteers at the SKYWARN Amateur Radio Station at One Time

No more than three (3) SKYWARN volunteers should be in the forecast office at one time. If, for some reason, there are more than (3) volunteers at the NWS, please take shifts. Off-duty Amateurs may monitor other nets from the NWS lunch room or from their cars in the parking lot.

4.1.4 Preparations for Extended Activation

SKYWARN amateur radio volunteers should be prepared for an extended stay at the NWS if SKYWARN is activated for a hurricane or for severe winter long-duration storms. The nearest food store is about one (1) miles from the forecast office and may not be open or accessible during extremely severe weather. Volunteers are responsible for bringing food, medications and personal hygiene supplies to maintain themselves for the duration of their stay at the NWS. Please be prepared to be as self-sufficient as possible.

4.2 How To Volunteer for Duty at the NWS

DO NOT GO RUNNING TO THE NWS OR CALL THE NWS AT THE FIRST SIGN OF BAD WEATHER. To be a well coordinated and effective operation we must follow protocol:

1. NWS determines a need for SKYWARN activation and activates SKYWARN.

2. The NWS will contact the SKYWARN Regional Coordinator to notify them that SKYWARN has been activated and to have them arrange for volunteer operators to staff the NWS SKYWARN station whenever 8 or more counties are in a Severe Thunderstorm and/or Tornado Watch.

Volunteer operators may also contact their county SKYWARN EC, or his/her designate, to inform them of their availability. Do not be insulted if your services are not needed at that time. As the weather situation changes, staffing needs may also change.

4.3 Interaction With The Forecasters

The forecaster who briefs the NWS SKYWARN operator upon arrival at the NWS will most likely be the contact person until the NWS shift changes. Please follow your instincts on how to pass information to the forecasters. If the information is CRITICAL and POTENTIALLY LIFE THREATENING, bring this information to the forecaster IMMEDIATELY, otherwise you will need to gauge the situation as to whether the information is important enough to bring to the forecasters attention immediately or if it can wait five or ten minutes until a forecaster comes to the amateur radio station as part of his or her duties. It is a delicate balance to make this critical part of the operation successful and it must be handled with discretion, tact and diplomacy by the operators.

4.4 Briefing Upon Arrival at the NWS

Upon arrival at the NWS, immediately identify yourself to a forecaster as a SKYWARN amateur radio operator and ask the forecaster for a briefing on the severe weather situation. You should get the following information from the forecaster:

1. Where storm(s) are located and in which direction(s) they are traveling;

2. Characteristics and history of the storm(s) (i.e., hail, damaging winds, tornadoes, snow, etc.);

3. What geographic locations/counties are of primary concern to the forecasters; and,

4. The latest severe and/or special weather statement(s) to be read over the net.

4.5 Initial Setup

After receiving the briefing, the SKYWARN operator should take the following steps:

1. Size up the situation and make a plan of attack;

2. Get sufficient copies of the action log forms. Please date the sheet and write legibly. The reports may be used for Storm Data (an official record of the event).

3. Find pens and pencils on the adjacent desk.

4. Set up Radio #1 (VHF/UHF) to access the county(s) of primary concern and roving nets both on two meters and 440. Radio #2 (HF) will require at minimum a General Class license. This transceiver will be utilized for liaison with state agencies and possibly direct communications with outlying areas when other means have failed. The primary and secondary frequencies for each county are listed as Appendix C.

5. Take another deep breath and start roving for reports.

6. Keep the forecasters informed of the reports received.

The NWS SKYWARN volunteer will act as a liaison between the nets and the forecasters. The NWS SKYWARN operator will record all information from the net on the SKYWARN severe weather reporting sheets, will break into the net and get further details (fills) from reporting stations as needed, will read severe and special weather statements over the air, when available, and will interface with the forecasters and inform the net of special areas of interest to the forecasters.

4.6 Ending SKYWARN Operations

When the severe weather situation calms down, a forecaster will indicate to the SKYWARN operator of the NWS station that the station can be secured. At that time, the operator should perform the following shutdown steps:

1. Ask for any additional reports of severe weather.

2. Notify all county nets in operation that the SKYWARN operations are ending and that any further reports of severe weather must be telephoned into the NWS.

3. The station MUST be left in a clean condition READY for the next activation.

4. Please staple all reports and statements together and hand them to the lead forecaster!

V. INTERFACING WITH OTHER GROUPS

The NWS is often asked to communicate with other groups in addition to amateur radio operators. Every attempt should be made to have the broadest possible inclusion into the SKYWARN Net. At the present time, there are no facilities to monitor citizens band frequencies at the SKYWARN Amateur Radio Station and no such facilities are planned. Accordingly, if groups such as REACT would like is participate in SKYWARN nets, it is imperative that the group coordinate with one of their members who is also an amateur radio operator who can relay the reports.

VI. SKYWARN HF OPERATIONS

HF will be set up at NWS for use as backup during major communications outages, contact with government agencies during widespread events. It will also serve as the primary means of communications with areas that are not reachable on VHF, as well as for MARS and SHARES.

VI. PUBLICITY and PUBLIC RELATIONS

An important facet of SKYWARN operations is public relations. SKYWARN provides ample opportunities to demonstrate the unique capabilities of amateur radio as well as the volunteer and public spirit of amateur radio operators. Any questions from the media should be directed to the NWS Severe Weather Coordinator on duty.

There are constant challenges to the radio frequencies set aside for amateur radio use. SKYWARN provides an identifiable and extremely visible opportunity for pursing amateur radio in its best light. Severe weather is always of interest to the media. As a direct result of SKYWARN activities, generally, and SKYWARN participation in training exercises, as well as region wide communication drills in particular, and demonstrated professionalism and results as noted in NWS weather statements and reports on severe weather events, a number of Emergency Operations Centers have recognized the benefits of having amateur radio capabilities at their disposal. Thus, the SKYWARN program not only benefits the NWS and the public, but also helps to preserve amateur radio as a national resource.

The NWS does its best to promote the capabilities of the SKYWARN amateur radio net. NWS works closely with FEMA, the American Red Cross, the FCC in Emergency Broadcast Communications and with numerous state and local emergency management agencies. Therefore, SKYWARN has been, and will continue to be, an important vehicle to showcase amateur radio to the agencies involved in the allocation of privileges and frequencies.

SKYWARN has developed a large following of scanner enthusiasts, emergency managers and amateur radio operators. Please remember as you operate, members of the media are monitoring your communications. Let us continue to put Amateur Radio's "best foot forward."

MEMORANDUM OF UNDERSTANDING BETWEEN THE NATIONAL WEATHER SERVICE AND THE AMERICAN RADIO RELAY LEAGUE, INC.

I. PURPOSE

The purpose of this document is to state the terms of a mutual agreement (Memorandum of Understanding) between National Oceanic and Atmospheric Administration's (NOAA) National Weather Service (NWS) and the American Radio Relay League, Inc. (ARRL), that will serve as a framework within which volunteers of the ARRL may coordinate their services, facilities, and equipment with NWS in support of nationwide, state, and local early weather warning and emergency communications functions. It is intended, through joint coordination and exercise of the resources of ARRL, NWS, and Federal, State and local governments, to enhance the nationwide posture of early weather warning and readiness for any conceivable weather emergency.

II. RECOGNITION

The National Weather Service recognizes that the ARRL is the principal organization representing the interests of more than 690,000 U.S. radio amateurs. Because of its field organization of trained and experienced communications experts, Amateur Radio Service volunteers can be of valuable assistance in early severe weather warning and tornado spotting.

ARRL recognizes the National Weather Service's statutory responsibility to provide the following meteorological services for the people of the United States:

1. NOAA's National Weather Service provides weather, hydrologic, and climate forecasts and warnings for the United States, its territories, adjacent waters and ocean areas, for the protection of life and property and the enhancement of the national economy; and,

2. NWS data and products form a national information database and infrastructure which can be used by other governmental agencies, the private sector, the public, and the global community.

III. ORGANIZATION OF THE AMERICAN RADIO RELAY LEAGUE

ARRL is a noncommercial membership organization of radio amateurs, organized for the promotion of interest in Amateur Radio communication and experimentation, for the establishment of networks to provide communications in the event of disasters or other emergencies, for the advancement of the radio art and of the public welfare, for the representation of the radio amateur in legislative matters, and the maintenance of fraternalism and a high standard of conduct. A primary responsibility of the Amateur Radio Service, as established by the Federal Communications Commission, is the rendering of public service communications for the general public, particularly in times of emergency. Using Amateur Radio operators in the amateur frequency bands, the ARRL has been serving the public, both directly and through government and relief agencies, for more than ninety years. To that end, the League created the Amateur Radio

Emergency Service ® (ARES) ® and the National Traffic System (NTS). The League's Field Organization consists of seventy-one administrative sections managed by elected Section Managers. A Section is a League-created political boundary roughly equivalent to states (or portions thereof). The Section Manager appoints expert assistants to administer the various emergency communications and public service programs in the section. Each section has a vast cadre of volunteer appointees to perform the work of Amateur Radio at the local level, under the supervision of the Section Manager and his/her assistants.

IV. ORGANIZATION OF THE NATIONAL WEATHER SERVICE

National Oceanic and Atmospheric Administration's (NOAA) National Weather Service consists of 122 weather forecast offices, 13 river forecast centers, 9 national centers, and other support offices. NWS scientists provide weather, water, and climate forecasts and warnings for the United States for the protection of life and property, and the enhancement of the national economy. The NWS' national headquarters is located in Washington, D.C., and there are six regional headquarters: Eastern, Southern, Central, Western, Alaska, and Pacific.

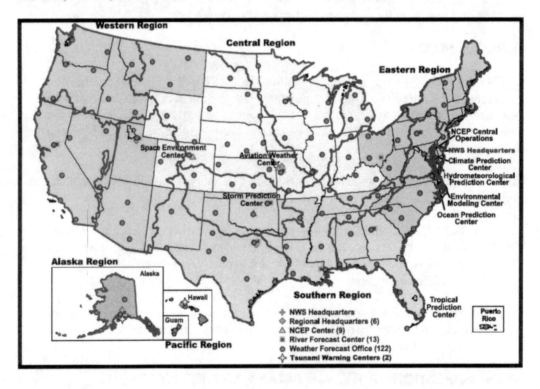

Skywarn® is the National Weather Service's severe weather spotting program. Radio amateurs have assisted as communicators and spotters since the program's inception in the late 1960s. In areas where tornadoes and other severe weather have been known to threaten, NWS recruits volunteers, and trains them in proper weather spotting procedures. These dedicated citizens help keep their local community safe by conveying severe weather reports to their local NWS Forecast Office. Skywarn spotters are integral to the success of our Nation's severe weather warning system.

Warning Coordination Meteorologists (WCMs) serve as the NWS' principal liaison with its customers and partners in the evaluation and improvement of its products and services. WCMs are responsible for maintaining the working partnership with the local ARRL Skywarn organizations. There are 132 NWS Warning Coordination Meteorologists (WCMs) located throughout the country: 122 Weather Forecast Offices, 6 Regional Headquarters, National Headquarters, the Storm Prediction Center, the National Hurricane Center, and the National Aviation Weather Center.

V. PRINCIPLES OF COOPERATION

A. ARRL agrees to encourage its volunteer Field Organization appointees, especially the Amateur Radio Emergency Service, to contact and cooperate with National Weather Service Warning Coordination Meteorologists for the purpose of establishing organized Skywarn networks with radio amateurs serving as communicators and spotters, consistent with rules and regulations of the Federal Communications Commission.

B. ARRL further agrees to encourage its Section management teams to provide specialized communications and observation support on an as-needed basis for NWS offices in other weather emergencies such as hurricanes, snow and heavy rain storms, and other severe weather situations.

C. The National Weather Service agrees to work with ARRL Section Amateur Radio Emergency Service volunteers to establish Skywarn networks, and/or other weather emergency alert and relief systems. The principal point of contact between the ARRL Section and local NWS offices are the Warning Coordination Meteorologists. Local Warning Coordination Meteorologist contact information is available at: www.stormready.noaa.gov/contact.htm. Contact information for ARRL Section volunteer leaders is available at www.arrl.org/sections. The national contact for ARRL is the Emergency Preparedness Manager at ARRL Headquarters, Newington, CT 06111. The national contact for NWS Warning Coordination Meteorologists is the Office of Climate, Weather and Water Services, WCM Program, 1325 East-West Highway, Silver Spring, MD 20910.

Kay Craigie, N3KN   Date May 12, 2011
Kay Craigie, N3KN
President, American Radio Relay League, inc.

David B. Caldwell   Date 6/9/2011

Printed Name: David B. Caldwell

Title: OCWWS Director

NOAA National Weather Service

Appendix 5

False Statement Notice

24.00 *FALSE STATEMENTS*

24.01 *STATUTORY LANGUAGE: 18 U.S.C. § 1001*

§1001. ***Statements or entries generally***

(a) . . . [W]hoever, in any matter within the jurisdiction of the executive, legislative, or judicial branch of the Government of the United States,[1] knowingly and willfully —

> (1) falsifies, conceals or covers up by any trick, scheme, or device a material fact;
>
> (2) makes any materially false, fictitious, or fraudulent statements or representation; or
>
> (3) makes or uses any false writing or document knowing the same to contain any materially false, fictitious, or fraudulent statement or entry;

shall be fined under this title or imprisoned not more than 5 years . . . .[2]

---

1. The ***False Statements Accountability Act of 1996***, Pub. L. No. 104-292, 110 Stat. 3459, changed the language of Section 1001, which previously criminalized false statements made "in any matter within the jurisdiction of any department or agency of the United States . . . [.]" The ***False Statements Accountability Act*** superseded the Supreme Court's 1995 decision in ***Hubbard v. United States***, 514 U.S. 695, 702-03 (1995), which held that the previous version of Section 1001 prohibited only false statements made to the executive branch. The ***False Statements Accountability Act*** extended the application of Section 1001 to false statements or entries on any matter within the jurisdiction of the executive, legislative or judicial branch of the federal government. However, this prohibition does not apply to a party to a judicial proceeding, or to that party's counsel, "for statements, representations, writings or documents submitted by such party or counsel to a judge or magistrate in that proceeding." 18 U.S.C. § 1001(b).

2. The ***Intelligence Reform and Terrorism Prevention Act of 2004***, Pub. L. No. 108-458, 118 Stat. 3638, with an effective date of December 17, 2004, increased the penalties under Section 1001 for crimes involving international or domestic terrorism to include a term of imprisonment
of not more than 8 years. Two separate pieces of legislation, each of which would increase the term of imprisonment under Section 1001 for crimes involving terrorism to not more than 10
years, are currently pending in Congress. See ***Counter-Terrorism and National Security Act of 2007***, H.R. 3147, 110th Cong. (1st Sess. 2007); ***Violent Crime Control Act of 2007***, H.R. 3156, 110th Cong. (1st Sess. 2007).

Under 18 U.S.C. § 3571, the maximum fine under Section 1001 is at least $250,000 for individuals and $500,000 for corporations. Alternatively, if any person derives pecuniary gain from the offense, or if the offense results in a pecuniary loss to a person other than the defendant, the defendant may be fined not more than the greater of twice the gross gain or twice the gross loss.

The purpose of Section 1001 is "to protect the authorized functions of governmental departments and agencies from the perversion which might result from" concealment of material facts and from false material representations.

Appendix 6

Integrating Google Earth, NWS Data and APRS Using KML

By Robert Andrews, KØRDA

I wanted to find an easy way to integrate mapping, radar data, and APRS tracking data for SKYWARN events. I had been told to look at software from *GRLevelX*. Using a combination of pay software and add-ons, it appears possible. However, I wanted a way to do it for free. And I found it!

In this section, I am going describe the process I followed to:

- Get a mapping program
- Show watch and warning data
- Show local and national radar data
- Show APRS tracking data
- All for *free!*

FIRST, GET A MAPPING PROGRAM

To start, download and install the *Google Earth* program. Once installed, launch the program. When you launch the program, you will see the screen is divided into two sections. On the left is your working panels and on the right is the large viewing area. Your left panels are divided into three sections: Search, Places, and Layers. We will be using all three of these (**Fig 1**).

Once *Google Earth* is open, we want to set some basic settings. In the Layers panel (**Fig 2**), I recommend showing the following items:

- Roads
- 3D Buildings
- Borders and Labels
- Terrain

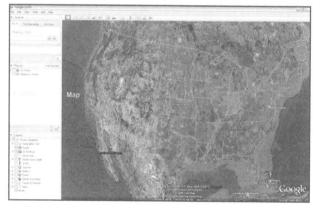

Fig 1

Fig 2

However, the choice is up to you to show more or less detail. Now, the observant person may see an option for Weather right there. I do not like this option because they are pulling their data from Weather.com. They also only provide a national composite radar image. This means that Weather.com needs to grab the data from the National Weather Service (NWS) and stitch it all together into a national map image. This causes both a slight delay and a loss of some precision.

Integrating Mapping and Weather Data   A6.1

Now that we have the map showing the features we want, use the Search panel to get zoomed in on your county. I recommend adjusting the zoom to show your county and about one county more around it on all sides. This gives you a good view of what is coming your way. Once you have the zoom adjusted, click on the View menu and select "Make this my start location." By doing this, *Google Earth* will automatically go to this view when the program is opened saving you the need to search for your location each time (**Fig 3**).

Fig 3

ENTER KML

So at this point, we have met the first bullet of our requirements. But how can we get data directly from the NWS or APRS data into *Google Earth*? It is done through the magic known as *Keyhole Markup Language (KML)*. KML is an extension of the *eXtensible Markup Language (XML)*. KML allows for the overlay of images, data points, polygons, icons and links to inline rich-text data, HTML, or external Web sites. Often, KML files will be compressed or zipped into a KMZ file. (Note to nerds: you can download a .KMZ file, rename it .ZIP, and extract the .KML file from it. You can then rename the .KML file to .XML and open it to see the raw data.) Keep in mind that what you see may simply be a link to Web data. We will see two examples of this as we go down our shopping list of wants.

ADD WATCH AND WARNING DATA

Our next set of requirements requires data from the National Weather Service. All the data that we want to use is already provided. And now, the NWS provides this data via KML format! And to make it super easy, they created a KML / KMZ generator page that puts all items in one simple-to-use page (**Fig 4**).

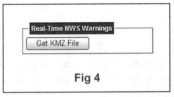

Fig 4

The second bullet is to see current watches and warnings. This is crucial to see what current hot spots exist. To add this data to *Google Earth*, go to the KMZ Generator page and click the "Get KMZ File" button for Real-Time NWS Warnings. Click the Open button when prompted. *Google Earth* will launch if it is not already open (**Fig 5**).

In the Temporary Places section of the Places panel, we see the NWS Warnings item is added (**Fig 6**). We expand this to see there are two items under NWS

Fig 5

Warnings. The first is a folder called National Weather Service. Notice the standard folder icon showing this contains static information. Inside the National Weather Service folder are three items: National Weather Service, NOAA and Warning Legend. The first two add logos to the screen while the third adds the warning legend: Red for Tornado, Yellow for Severe Thunderstorm, Green for Flash Flood, and Blue for Marine. I don't particularly like parts of my screen getting covered by the NWS and NOAA logos. To get rid of them, you can simply right click on those items and select delete.

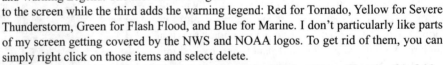

Fig 6

The second folder under NWS Warnings is a little different. You will notice this folder has a little network line under it. This tells us that the content of this folder is not static and is actually pulled from the Web live. If we right click on the NWS Warning Products and select Properties, we can see what is really happening. Right on top, we see the link to **http://www.srh.noaa.gov/data/radar/poly.kml** We start to see the nesting capabilities of KML. The initial KML file contained static items as well as this Web link. Therefore every time we launch *Google Earth*, it will pull in the current information. (Note to nerds: you can directly put this URL into your browser, save the .KML file, rename to .XML, and open it to see the raw data.)

When we first selected this item, it was labeled "Real-Time". We can review the refresh details for this network link. We can see here that it is set to pull new data every minute. So while it is labeled as real time, we are actually doing an update every minute. If you watch your screen closely and look at that network folder icon, you will see an animated folder icon as it is updating from the Web. You will see 'data' flowing over that little network line on the bottom of the folder (**Fig 7**).

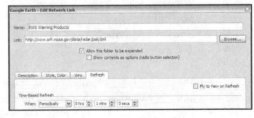

Fig 7

When we look at the map and see a warning, we can click on the little icon associated with the warning polygon. In this example, when I click on the yellow lightning bolt box on the upper right of the yellow warning area, I get a pop-up box right in *Google Earth* with the complete details of the warning, sightings, times, and locations (**Fig 8**).

Right now, our custom version of the NWS Warnings folder is stored in Temporary Places. As the name implies, when you close *Google Earth*, this will be lost. To save this, drag and drop the NWS Warnings folder from Temporary Places to My Places. Then click on the File menu, select Save, and Save My Places. Inside the My Places folder will also be your Starting Location saved from above.

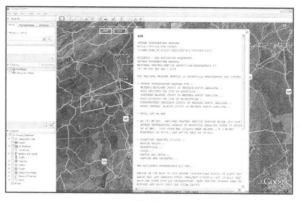

Fig 8

Fig 9

If you wanted to remove this, right click and delete it or just resave your starting location using the method above (**Fig 9**).

At the front of each folder and item is a check box. When you check something at a higher level, all sub-levels and items will be checked. Therefore, if you did not want to delete those logos, you could simply uncheck them. However, for the ease of just being able to click the collapsed NWS Warnings folder and have all the contents selected and shown, it is great. If you want to see what each element does, use the individual check boxes to turn each item on and off.

ADD RADAR DATA

On to the next bullet: radar images. For my personal preferences, I actually wanted both local and national radar data. I find the local radar image has a little more detail than the national image. From the local radar site, NWS actually provides several types of radar images. Because of this, I actually create three different radar image items. For quick access, the first one I add is the Short-Range Reflectivity from the La Crosse site (**Fig 10**).

Next, I add the All Images for the La Crosse site. This gives me a folder that contains each of the seven different images they offer. Finally, I add the Lower 48 States (CONUS) National mosaic reflectivity image (**Fig 11**).

As you may have seen before when you were deleting items, you can also rename them. So, after deleting unwanted elements, moving them from Temporary Places to My Places, reorganizing, and renaming, I get a structure that looks like the image in **Fig 12**. This allows me to quickly select which items I want to see.

Fig 10

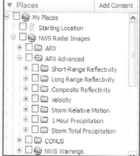

Fig 12

When I turn on the Continental US data, I can see an overlay for the entire county and where any action is. Since the CONUS image requires building the image for the entire US, it will overlay with the local radar image. I recommend using only one or the other at a time, as they may not line up exactly over each other. At the time I am creating this document, there is no activity in the local region, so my La Crosse (ARX) radar images doesn't show anything. However, in the Northeast, there is activity. **Fig 13** is a sample of that part of the country, showing the radar and warnings overlaid on the map.

Fig 13

ADD APRS DATA

So now we have a great map tool and have integrated all the weather data we want from the NWS. The next step is to add in the APRS data. Almost as easy as it was to add in the weather data, **APRS.fi** allows us to see the APRS data stream in KML form as well. Just go to **APRS.fi** and log in with your call sign. There is no profile or account; you can put in anything you want. On the right hand side of the map, under Other Views, you will see a link for *Google Earth*. Just click that and hit Open to bring APRS data into *Google Earth*.

When this item shows up in Temporary Places, it will load all the APRS locations for the past hour in the area of the map you are looking at. This will include station locations, paths, and weather data. Again, we can notice that this is a network folder, by its icon, and not static content. This means it is talking to **APRS.fi** to get the live data. It is actually smart enough to know what region of the map you are zoomed to. As you re-zoom or move around the map, it will trigger a refresh of the data and it will pull in the APRS data for your current view. It will also refresh every five minutes if the view does not change. (Note to nerds: if you want to see how this works, right click and select Properties, as I showed above.) Due to the large volumes of APRS data, it will limit it to 1000 items on the screen. So if you pull out to show too large an area, such that greater than 1000 items would need to be drawn, it will display only some of them. A message will show on the screen indicating that the number of items was limited and to zoom in to reduce the area to see all items.

In addition to general APRS traffic from the past hour, you can track specific objects. After logging into **APRS.fi**, put the call sign and *SSID* (specific number for users that have multiple APRS devices) into the Track text box and hit search. This will show just that item on the **APRS.fi** map. However, now when you click the *Google Earth* link, in addition to the general APRS traffic folder for the last hour, you will get a second network folder for that particular call sign. It will show the last known location and path for that person, even if it was not during the past hour. If you have multiple calls you want to track, search for them on the site, add them to *Google Earth*, and move just those tracking folders from Temporary Places to My Places. As a reminder, don't forget to save your My Places. **Fig 14** shows how I normally keep my personal My Places arranged. The Starting Location does not need to be checked. It will automatically set the map to that position. By checking that box, it will only add a "Starting Location" label to your map, which may get in the way of other data. I can quickly add in my local or national radar as well as watches and warnings. Finally, I have check boxes to add in all APRS data in my view as well as track my specific items.

Fig 14

So that completes the list of requirements! We now have a map program, with live data from the NWS showing both warnings and radar data. We have direct access to the details on those warnings. We can see both local and national radar. And we even integrated APRS data and paths onto our map. So what does this look like when all combined? Take a look at **Fig 15**. As you can see from this sample, having all this information together makes a very powerful tool for spotters or anyone at all. And don't forget the best part: all of this is *free!*

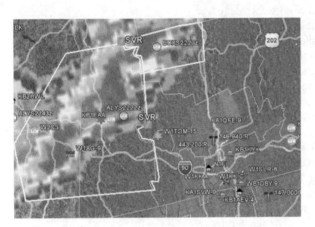

Fig 15

NWS LOOP CAPABILITIES

Recently the NWS has updated their loop capabilities. In the past, loops were hard coded to a specific time period. This means that when you added the loop, it worked for that time. However, if you came back a week later, the loop was coded to look at the original time one week ago. Now, the NWS has updated their loop code to always update in real time. This means that if I add a two-hour radar loop to my list of places, when a new 'frame' is added (every 10 minutes), it will pick it up instantly. This means that your loop is never more than two hours ten minutes old. Plus, if you save, close, and reopen, it will pull in the current time's loop. Depending on the loop you choose, it may be a one-hour or two-hour loop. The larger the coverage area, the shorter the loop and *vice versa*; smaller coverage areas give longer loops. The only thing they do not have that I would like to see is a full CONUS loop.

So how do you get these new loops? Just use the same KMZ Generator page to access the loop sections. For a single radar site, use the Single Image Loop section. For wider coverage, use the Regional Mosaic Loop section (**Fig 16**).

Because of these enhancements, I have added them into my standard saved locations (**Fig 17**). You can see in my new panel I have two loops in addition to everything I had above. I have added in a single image loop for my closest radar site.

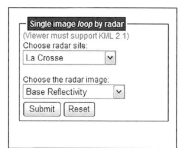

Fig 16

This gives me a two-hour loop. I also added the regional multi-radar site loop for me, the Upper Mississippi Valley. This gives me a one-hour loop. (Note to nerds: for those of you looking close, you can see how this new loop set-up works. Take a close look at the folder icons for the loops. You can see that they have the data cable icon showing that they are network folders. This means they do not store any static info but are always pulling from the Internet.) I will leave it up to you if you want to explore the properties of these new folders using the methods described above.

(Figs 1- 17: Source *Google Earth*. Used with permission.)

Fig 17

Appendix 7

WX4NHC: Amateur Radio Station at the National Hurricane Center

AMATEUR RADIO STATION AT THE NATIONAL HURRICANE CENTER

WX4NHC 2020 - Who we are and What we do
Our 40th year of Volunteer Public Service at NHC

Our Mission – to help save lives

WX4NHC is an Amateur Radio Station, also known as "Ham Radio", located at the National Hurricane Center in Miami, Florida. The station has been totally assembled from donated equipment and is operated by an organized group of volunteer amateur radio operators since 1980.

WX4NHC has been activated whenever a hurricane is within 300 miles of land fall in the areas of the western Atlantic, the Caribbean or the eastern Pacific. We also provide Emergency Backup Communications from NHC to NWS Offices and other agencies in case of local landfall.

The WX4NHC Team is composed of 20 specially trained volunteer operators that operate the Ham Radio station in 3-hour shifts. For example: during the Historic 2005 Hurricane Season the station was on-the-air, sometimes with two to three operators at a time, for more than 500 hours. About 400 Surface Reports were received during Hurricane/Super Storm Sandy. We operated twice when NHC was inside the Eye of a hurricane (Katrina and Wilma) and collect hundreds of reports each hurricane season!

These "**Surface Reports**" provide the forecasters with supplemental weather and damage data that are not normally available to them and are frequently incorporated into their advisories as they provide a human perspective and Eye Witness accounts of what people are experiencing during a hurricane.

The WX4NHC operators work in conjunction with the **Hurricane Watch Net, VoIP WX-Talk Hurricane Net** and other volunteer networks to collect real-time "Surface Reports" for the NHC Hurricane Specialists via Amateur Radio using many modes such as HF "Shortwave" Radio, VHF/UHF Radio, VoIP (Voice over Internet Protocol) systems; EchoLink and IRLP, Automatic Packet Reporting System (APRS) and volunteer weather observer networks, ON-NHC (Observers Network) & CWOP (Citizen's Weather Observers Program), using our on-line report form, email and Fax. WX4NHC also receives Surface Reports via Winlink, DSTAR, DMR and modes listed on our website. WX4NHC also relays Hurricane Advisories via the Ham Radio Nets to the hurricane affected areas and governments when conventional communications have been interrupted.

The WX4NHC Team has been nationally recognized for their volunteer international humanitarian efforts by the National Hurricane Conference and the South Florida Hurricane Conference.

NHC and the WX4NHC Team is very grateful for the participation of volunteer Ham Radio Operators before, during and after hurricanes. Whether you are directly affected by the hurricane or a distant relay station, you are an important part of the communications link that provides NHC with those important eye witness surface reports.

Without your efforts to communicate those hurricane reports, WX4NHC would only be listening to static.

THANK YOU!
www.wx4nhc.org

AMATEUR RADIO STATION
AT THE
NATIONAL HURRICANE CENTER
11691 S.W. 17 STREET   MIAMI, FLORIDA   33165

How to Contact Amateur Radio Station WX4NHC

Amateur Radio HF Frequencies - (single sideband mode)
20 meters : **14.325 MHz** Hurricane Watch Net (Main frequency during Hurricanes)
40 meters : 7.268 MHz Hurricane Watch Net (secondary frequency), shared Water Way Net
80 meters : 3.815 MHz Caribbean Net, (Alternates: 3.950 : North Florida / 3.940 South Florida)

Amateur Radio EchoLink / IRLP
EchoLink VoIP Hurricane Net, Conference: **"WX-TALK"** (Node 7203)
IRLP Node 9219

Amateur Radio VHF/UHF Frequencies
VHF : 147.470 MHz simplex - Coordination frequency for NHC ops. *(official use only, please)*
VHF : 147.000/146.400 repeater (PL 94.8 Hz), 146.925 backup repeater (PL 94.8 Hz)
UHF : 444.600/449.600 repeater (Local use South Florida PL 94.8 Hz)
UHF : 444.600/449.600 repeater, SARNET (During Florida Hurricanes Statewide PL 167.9 Hz)

APRS mode Frequencies
HF : 30 meters : 10.151 MHz (LSB)
VHF : 2 meters : 144.390 MHz simplex

Internet Home Page : www.wx4nhc.org
Online Hurricane Weather Report Form : www.wx4nhc.org/WX-form1.php
  WX4NHC Email : wx4nhc@wx4nhc.org

Amateur Radio Coordinator:
John McHugh, K4AG Email: k4ag@arrl.net
Asst. Amateur Radio Coordinator:
Julio Ripoll, WD4R Email: wd4r@arrl.net

AMATEUR RADIO STATION – WX4NHC
AT THE
NATIONAL HURRICANE CENTER

WEATHER SURFACE REPORT

Hurricane:

Observation Station Call Sign (or full name):

Geographic Location (city, state, country):

Distance from ocean: **Height above sea level:**

GPS Location: (Latitude/Longitude): _____ North / _____ West

Date: Observation Time: ☐ GMT   ☐ LOCAL

WIND SPEED: _____   ☐ MPH   ☐ KNOTS/H   ☐ KILOMTR/H
(SUSTAINED ONE MINUTE)   ☐ ESTIMATED   ☐ MEASURED
GUST WIND SPEED: _____   ☐ MPH   ☐ KNOTS/H   ☐ KILOMTR/H
WIND DIRECTION: _____

Barometric Pressure: _____   ☐ INCHES   ☐ MILLIBARS
CALIBRATED TO SEA LEVEL? _____   ☐ FALLING   ☐ RISING

Surface Observations: *Flooding, sea level storm surge, Rain amount, Damage Report.*

_____
_____
_____
_____
_____
_____
_____
_____
_____
_____

WX4NHC Operator on duty: _____

Relayed by: ☐ HWN   ☐ VOIP   ☐ APRS   ☐ ON-LINE/EMAIL   ☐ OTHER:_____

Our Mission – to help save lives
www.wx4nhc.org

AMATEUR RADIO STATION AT THE NATIONAL HURRICANE CENTER

NHC Directors express the importance of Amateur Radio Communications.

Max Mayfield, former Director of the National Hurricane Center:
"Ham Radio Operators can give us information that we cannot get from other sources. These surface reports are valuable to us, as they give us information of what is actually happening on the ground. We at NHC appreciate your participation in relaying surface reports from hurricane affected areas, as well as distributing the Hurricane Advisories to those with no other means of receiving these vital warnings."

Bill Read, former Director of the National Hurricane Center:
"In an era with increasing reliance on high speed technology, we still need the capability to relay critical information to and receive critical information from, those communities in their time of greatest need - in times of disaster when most technology has failed. Frequently the only viable form of communication are the dedicated HAM radio operators in or near the disaster area. We at NHC are grateful for the support of the team of radio operators staffing WX4NHC during Tropical Cyclone events and assisting in potentially lifesaving communications."

Richard Knabb, Director of the National Hurricane Center:
"When I was a hurricane specialist here at NHC, especially during the extremely busy year of 2005, I frequently relied on information from dedicated HAM radio operators in the U.S. and in many other countries. They are key partners with us as we disseminate forecasts and warnings and collect all available data both while an active tropical cyclone is out there, and after the event when the crucial task of documenting the impacts is conducted. Our HAM radio friends are as passionate as we are at NHC about disaster safety and preparedness, and they provide a method of communicating that has withstood the test of time, even in the midst of other technological advances. Thank you to all who participate in and support this important community."

Ed Rappaport, Deputy Director of the National Hurricane Center:
"Because there are very few "official" weather stations, the NHC knows the information you provide could be critical to its forecasters, and to the people in your community and to those along the future track of a tropical storm or hurricane. This has been the case many times in the past. In Hurricane Andrew, for example, one special observer provided the highest known surface wind speed for that storm (177 mph), while others observed a central pressure (922 mb or 27.2 inches) that was 10 mb lower than what was expected from the official source (even from reconnaissance aircraft !). More recently, the highest wind speeds noted in North Carolina during Hurricane Fran came from observers like you.
This data is also important to the NHC even if it can't be sent immediately. The NHC reanalyzes every storm after it ends and very often revises its earlier estimates. The data form found on the WX4NHC main menu shows the kind of data that is the most important to the NHC. From previous storms, the NHC, and its sister agency the Hurricane Research Division, have found that several considerations can help improve the reliability of the data.
Thank you. We look forward to hearing from you!"

www.wx4nhc.org

Appendix 8

Lightning Protection for the Amateur Radio Operator's Home

Lightning protection is a safe investment for any station.

Jennifer Morgan and Michael Chusid

It is every radio operator's worst nightmare. Millions of volts suddenly leap out of the sky, striking your home, antenna, or other conductive surfaces. Billions of watts race down transmission lines or through the building's structure, destroying your transmitter, amplifier, receiver, or other elements of your radio apparatus. Your prized station is now just trash. And if you happen to be online when the "signal" comes in, you could be toast, too!

Congratulations — you have just become a victim of one of the approximately 25 million lightning strikes that occur in the United States each year. You have learned the hard way that in the contest between lightning and your equipment, lightning usually wins.

An excellent, three-part article on "Lightning Protection for the Amateur Radio Station," by Ron Block, KB2UYT (now NR2B), was published in the June, July, and August 2002 issues of *QST* and provides sound guidelines for protecting your station.[1] But as stated in Part 1 of the series, the goal is "to establish a 'zone of protection' within the radio room, as opposed to the whole house or building." The particulars of protecting the unique electronics in a radio setup are beyond the scope of this article, and a specialist may need to be consulted. This article recommends that the protection of your equipment begins with protecting a seemingly minor accessory to your station — *your home*.

Special Risks for the Amateur

Lightning strikes pose special risks for the Amateur Radio operator. To increase signal coverage, operators often place their antennas as high as possible — for example, on towers or rooftops — increasing the risk of being at the receiving end of a strike.

When an enthusiastic amateur upgrades their station, he or she inadvertently becomes more vulnerable. That's because, as radio equipment improves, electronic circuits become miniaturized and, thereby, more susceptible to damage from energy surges.

Amateur Radio operators take pride in being of service during emergencies. Unfortunately, lightning strikes occur during hurricanes, tornados, forest fires, floods, blizzards, and other extreme weather events. Emergencies are the exact time that Amateur Radio operators are needed the most, and the worst time to discover latent damage or degradation.

Unfortunately, several myths about lightning protection come into play.

Myth 1: I don't need a lightning protection system (LPS) because lightning strikes are rare.

According to the Insurance Information Institute, claims for lightning losses cost nearly a billion dollars in

Figure 1 — In this drawing, taken from an online lightning risk assessment, orange shows the footprint of a one-story house with a 10-foot eave height and a 14-foot ridge height. The brown area shows the "collection area" that contributes to the building's vulnerability to lightning. When a 30-foot-tall antenna is added to the lower left corner of the house (X), red shows increased size of the collection area due to the antenna's height. [Photo courtesy of East Coast Lightning Equipment, Inc.]

Figure 2 — The braided copper cable and large clamp are UL-listed for lightning protection and are much larger than the wire and clamp (near the top of the ½-inch-diameter grounding rod) required for ordinary household grounding. The two grounds are bonded to create a common ground point. [Photo courtesy of East Coast Lightning Equipment, Inc.]

2015. From 2010 to 2015, the average cost per claim rose 64%. These figures understate losses because much damage is not reported (or is not included because of deductibles), or is attributed to equipment malfunction instead of lightning. In addition, the frequency of lightning strikes appears to be increasing.

Keep in mind that low risk is not the same as no risk. Tennessee generally has a lower number of lightning strikes than Florida, but still had $24 million in homeowner lightning damage claims in 2015. Even in low-risk southern California, lightning storms caused deaths and injuries, set fire to homes, and knocked out power to Los Angeles International Airport — even at the height of last year's drought.

Myth 2: I don't need an LPS because my antenna is not the tallest structure around.

Lightning is going to go wherever it wants to. Lightning protection standards treat lightning as if it were a 300-foot-diameter sphere rolled across the surface of a building (see Figure 1). Any place the sphere contacts the building is a location where lightning can attach. The point of contact might be your antenna, but could just as readily be another point on the house. Moreover, lightning need not directly strike a structure to cause damage. The energy of a lightning strike can "side flash" from one object to another. Also, lightning can travel to your structure through metal objects (such as wire fences, plumbing, and cables) and even through the ground.

Myth 3: I don't need an LPS because I connected my antenna to a metal ground rod.

An LPS is more than just a wire to a ground rod (see Figure 2). The Insurance Information Institute cautions:

> Keep in mind lightning protection system design and installation is complex and not a do-it-yourself project. Installation is not typically within the scope of expertise held by general contractors, roofers, or even electricians, which is why the work is typically subcontracted out to specialists.[2]

Lightning Protection Fundamentals

Lightning typically begins within a cloud, where ice particles collide and generate static electricity. When the charge within the cloud grows sufficiently, the electrical insulating properties of air fail, and an ionized conductive channel to the opposite charge is established. The rapid discharge of electricity along this channel is lightning, both cloud-to-cloud and cloud-to-ground.

The power of a lightning strike is daunting. In 0.2 seconds, the air around the conductive channel can heat up to 50,000°F. An object in the path of a lightning strike may be subjected to as much as 3 million volts.

Lightning seeks the path of least resistance to ground. If the path is through your home, it could cause fire or structural damage. If it is through you, it could cause serious injury or death. And if it is through your equipment, you're likely to lose that equipment (see Figure 4).

Lightning protection systems work by creating an adequately sized, low-resistance path for lightning to flow around a structure into the Earth. An LPS should be designed and installed in compliance with the following standards (see Figure 3), based on technologies and principles that have been proven over the past 200 years:

■ National Fire Protection Association (NFPA) 780 — *Standard for the Installation of Lightning Protection Systems.*[3] UL (formerly known as Underwriters Laboratories) 96A — *Installation Requirements for Lightning Protection Systems* and 96 — *Lightning Protection Components*

■ Lightning Protection Institute (LPI) 175 — *Standard of Practice for the Design — Installation — Inspection of Lightning Protection Systems;* National Standard of Canada, CAN/CSA-B72-M87 — *Installation Code for Lightning Protection Systems* (applies to projects in Canada).

Figure 3 — Your station is not safe unless the building in which it is installed is also protected. A complete lightning protection system includes many components designed and installed to meet the standards of NFPA, UL, and LPI. [Photo courtesy of East Coast Lightning Equipment, Inc.]

Compliance with building codes governing normal electrical systems is not enough to protect against lightning. As stated in the International Association of Electrical Inspectors (IAEI) handbook, *Soares Book on Grounding and Bonding*, the "installation of a lightning protection system is much different from the installation of electrical service wiring." Furthermore:

> Specialized material and installation methods, such as that specified in NFPA 780 and UL 96, are required, and the system should only be installed by qualified personnel trained and certified in the installation of lightning protection systems.

Components

All components must be UL listed for lightning protection; UL-listed prod-

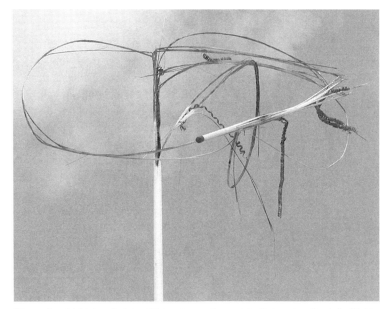

Figure 4 — Lightning destroyed this antenna. A properly designed and installed lightning protection system could have prevented this damage and provided protection for the owner's home and its contents.

ucts for electrical services are inadequate for the overwhelming force of lightning.

Components are made from high-grade aluminum, copper, or copper alloys due to their high electrical conductivity. Project conditions and adjacent building materials determine which of these metals can be used.

Air Terminals

Located at the top of a building, these are usually the first elements of an LPS that engage lightning strikes. Air terminals, colloquially known as lightning rods, can be as small as ⅜ inches in diameter by 10 inches tall. They are available in a variety of metals and styles to complement the design of your home.

Air terminals must be installed at the ends of roof ridges and corners of roof parapets, and at a maximum of 20-foot intervals. They must also be on high points, such as chimneys, antennas, and rooftop equipment (see Figure 5).

We caution against the use of so-called "early streamer emissions," "dissipation array," and "charge transfer" air terminals. Claims that these devices "attract" or "repel" lightning have been debunked by NFPA, court rulings, and international studies. The devices can function as individual air terminals if used in accordance with NFPA 780, but they do not reduce the quantity of air terminals needed to protect a building.[4]

Conductors

Made from multi-strand copper or aluminum cables, conductors connect the components of the LPS. The diameter of the cables is much larger than the wires used in home electrical service, allowing lightning to pass without sufficient resistance to generate heat (see Figure 6). This is important because cables may be near wood or other combustible materials in your house.

At least two widely separated down conductors are required for any structure. Larger structures and buildings with more complex rooflines will require more. An antenna tower or structural element can be used as a down conductor if it is made of metal at least 3/16 inches thick and is electrically continuous.

Bonding

No LPS is an island. Lightning flows through every conductive path to ground that it can find, including structural elements, piping, ductwork, coaxial cables, and other components. Even if these elements are grounded, they need to be interconnected to the LPS using appropriately sized bonding connectors. Otherwise, current will arc from one path to another with lower electrical potential.

Grounding

Every down conductor must connect to a grounding electrode. In most cases, copper-clad steel grounding rods are driven 10 feet into the ground, and must be located at least 2 feet outside of a structure's perimeter. Each grounding electrode must be interconnected with the other electrodes and with the building utility ground. Other grounding configurations, such as copper ground plates, may be required in dry sand and rocky soils or when building on rock (see Figure 7).

Surge Protective Devices

A surge protective device must be installed on every power, coax, signal, and other type of wiring entering your home. The devices must comply with NFPA 780 and, as applicable, either UL 1449 — *Standard for Surge Protective Devices* or UL 497 — *Standard for Protectors for Paired-Conductor Communications Circuits*. Installation is a job best performed by a licensed electrician. An Amateur Radio operator, however, can design and install a zone of protection to safeguard against transient surges.

Surge protective devices should be

Figure 5 — Existing buildings can be retrofitted with lightning protection systems. [Photo courtesy of Mr. Lightning]

Figure 6 — A professional lightning protection installer can show you how to reduce the cost of an LPS. Here, for example, a conductor from the roof is connected to a steel column that acts as the conductor to the floor level and an external ground system. [Photo courtesy of Priestley Lightning Protection, LLC]

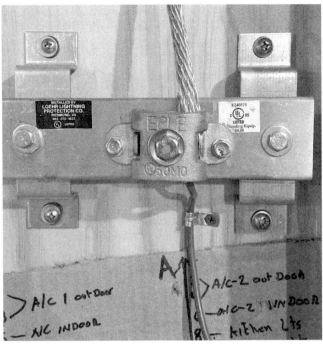

Figure 7 — This copper bar provides a convenient location to interconnect the grounds for various systems in the house. The braided copper cable is listed for use in lightning protection systems and leads to an interconnected ground system. [Photo courtesy of Loehr Lightning Protection Co.]

Figure 8 — Because much of a lightning protection system must be installed on rooftops or other high locations, it is advisable to have a trained professional with adequate protective gear do the work. [Photo courtesy of Labeled Lightning]

checked for damage at regular intervals.

Lightning Protection System Risk Evaluation

Not every building can justify the cost of an LPS. Fortunately, NFPA 780 has a "Simplified Risk Assessment" that you can use to evaluate your home's susceptibility to lightning damage. A qualified LPS designer can assist you with the risk assessment. Alternatively, you can run the calculations yourself using a free app, such as the one at **ecle.biz/riskcalculator**. It will ask you the following about your home:

- *Building size and height.* The size is based on the roof area of the building. The height includes the height of antennas, chimneys, and other rooftop equipment.
- *Frequency of lightning in your local area.* A map with this data is available at **ecle.biz/lightning-risk-map**.
- *Structures or trees near the building.* Structures that are taller than surrounding trees and buildings, or that are located on a hill, are at greater risk.
- *Structural and roofing materials.* Metal framing and roofing is less vulnerable to lightning damage than combustible materials are.
- *Value and combustibility of the building's contents.* In addition to your radio station, your home entertainment system, art collection, and other expensive features may justify a higher level of protection. Combustibility is another consideration; an empty metal shed will obviously need less protection than a wooden home owned by a Civil War reenactor who likes to store gun powder.
- *Occupants.* Risk increases if it will be difficult to evacuate children or other occupants.
- *Consequences of a strike.* If your family is willing to move out while lightning damage is being repaired, you are at a lower risk than if you want your equipment to be operable in an emergency.

Once risk factors are entered, the easy-to-use app performs the appropriate calculations to make a determination of your home's vulnerability to lightning. If the risk is greater than what is deemed the acceptable level of vulnerability, a building lightning protection system is recommended.

Buying and Maintaining an LPS

The IAEI recommends that only LPI-certified or UL-listed lightning protection specialists be contracted to design and install your LPS. Make sure your contract specifies that their work complies with the standards listed above (see Figure 8).

For extra confidence or to get lower rates from your homeowner's insurance provider, you may wish to have a third-party inspection service, such as the Lightning Protection Institute – Inspection Program (LPI-IP), inspect and issue a certificate that the LPS is installed correctly.

Once an effective system is in place, NFPA 780 recommends periodic maintenance to ensure that it stays in

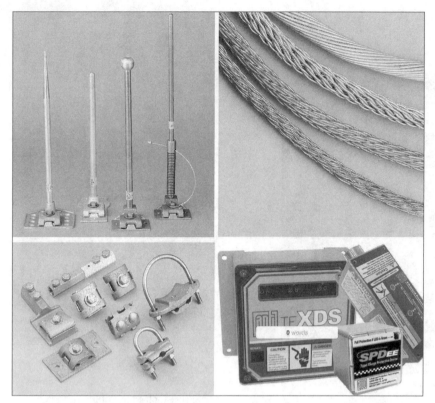

Figure 9 — LPS components must be UL-listed specifically for lightning protection. Clockwise from upper left: Air terminals, conductor cables, surge protective devices, and connectors. [Photos courtesy of East Coast Lightning Equipment, Inc.]

ARRL Field Day and Lightning Awareness

Across much of the US, Field Day is conducted in thunderstorm country. Every visitor to your Field Day site will appreciate the opportunity to learn about lightning detectors, especially with the eye-catching online maps at lighting detection websites like www.lightningmaps.org and en.blitzortung.org/live_lightning_maps.php?map=30. While they are looking at the maps, you can be building a lightning detector kit like this one from www.easternvoltageresearch.com/lightningdet10.html, or showing them how a commercial unit works to detect local lightning activity. (Detectors are available starting at about $60.) And you do have a National Weather Service radio, too, right? Tie it all together for your visitors and you'll be a thundering success!—Ward Silver, N0AX

compliance. Remodeling, reroofing, digging near the foundation, or the addition of antennas, satellite dishes, and rooftop equipment can compromise the efficiency of a lightning protection system, so it is worth having a lightning protection professional inspect the system on a regular basis. You may *believe* you are protected against lightning, but it is far better to *know*.

Don't Gamble with Lightning

A recent cost survey for professionally-installed LPS is available online at ecle.biz/coststudy. Antennas, extra surge protective devices (see Figure 9), and complex roof designs, will increase this estimate. Still, the cost of installing lightning protection on a typical home can cost less than many amateurs have invested in all the equipment, towers, and antennas required for their stations.

The chances of lightning strikes often increase during weather emergencies when Amateur Radio stations are most needed. Your station can help protect your fellow citizens, yet it is the lightning protection system over your head that protects *your* life and property. When designed and installed according to recognized standards, an LPS protects more than your radio station.

Notes
[1] www.arrl.org/lightning-protection
[2] bit.ly/insurance-press-release
[3] The current edition can be viewed at bit.ly/NFPA-780
[4] J. Morgan and M. Chusid, "Not all Lightning Protection is Created Equal," *Electrical Business*, Sept. 2016, bit.ly/NonConformingLightningProtection

Jennifer Morgan is co-owner of East Coast Lightning Equipment, Inc., the leading domestic producer of components for lightning protection systems. She can be reached through www.ecle.biz. Michael Chusid, an architect and a Fellow of the Construction Specifications Institute, is an authority on building products. Both are authorized by the Lightning Safety Alliance to present continuing education programs about lightning; see lightningsafetyalliance.org/education.html.

For updates to this article, see the *QST* Feedback page at www.arrl.org/feedback.

Notes

Notes

Notes

Notes

Notes

Notes

Notes

Notes

Index

Note: The letters "ff" after a page number indicate coverage of the indexed topic on succeeding pages. Page numbers starting with A refer to the Appendices. (For example, A6.1 is Appendix 6, page 1.)

A

AccuWeather: 3.18
    AccuPOP: 3.18
    Apps: 3.23
Advanced Hydrologic Prediction Service
    (AHPS Mobile): 3.17
Advanced Weather Interactive Processing
    (AWIPS): 4.4
After Action Report: 7.18ff
Allison House: 3.23
Amateur Radio Emergency Communications
    Course (ARRL): 4.2
Amateur Radio Emergency Service (ARES): ... 1.1, 1.4,
        4.1, 4.6, 7.1, 7.13ff
    ARES Form 1 - Public Service Activity Report: ... 7.16
    ARES Form 2 - Monthly DEC/EC Report: 7.16
    ARES Form 3 - EC Annual Report: 7.16
    ARES Form 4 - Monthly SEC Report to ARRL
        HQ: 7.16
American Meteorological Society: 4.4, 4.5
Antennas: 3.2
Apps: 3.21ff
    Top 10 Weather Apps: 3.22
ARRL
    Amateur Radio Emergency Communications
        Course: 4.2
    Amateur Radio Emergency Service
        (ARES): 1.1, 1.4, 4.1, 4.6, 7.1, 7.13ff
    ARES Form 1 - Public Service Activity Report: ... 7.16
    ARES Form 2 - Monthly DEC/EC Report: 7.16
    ARES Form 3 - EC Annual Report: 7.16
    ARES Form 4 - Monthly SEC Report to ARRL
        HQ: 7.16
    District Emergency Coordinator (DEC): 4.6, 7.17
    Emergency Coordinator (EC): 1.1, 4.6, 7.17
    Field Organization: 7.1, 7.17
    Memorandum of Understanding with the NWS: ..A4.1
    Report forms: 7.16
    Section Emergency Coordinator (SEC): 4.6, 7.17
    Section Manager (SM): 4.6
    Simulated Emergency Test (SET): 4.6
Automated External Defibrillator (AED): 4.3
Automated Surface Observing System (ASOS): 7.11
Automated Weather Observing System (AWOS): .. 7.11
Automatic Packet Reporting System
    (APRS): 3.4, 4.6, A6.1
Avalanche: 5.19
AWARE magazine: 4.4

B

Binoculars: 3.13
Books: 4.5, A1.1
Broadcastify: 3.3

C

C4FM: 3.5
Cameras: 3.10
    Image sharing: 3.12
    Video: 3.11
Cardiopulmonary resuscitation (CPR): 4.3
Cell phones: 3.9
Central Pacific Hurricane Center (CPHC): 3.17
Citizen Weather Observer Program
    (CWOP): 3.4, 3.23, 7.12
Climate Prediction Center: 4.4
Community Collaborative Rain, Hail and Snow
    Network (CoCoRaHS): 7.11
Community Emergency Response Team (CERT): ... 4.2
Cooperative Observer Program: 1.3

D

D-STAR: 3.5
    Severe weather use: 3.6
Damage reports: 7.16
Derecho: 5.6
Digital modes: 3.4ff
    Automatic Packet Reporting System (APRS): ... 3.4
    Echolink: 3.4
    Go-kit: 3.5
    Internet Radio Linking Project (IRLP): 3.4
Digital voice: 3.5
District Emergency Coordinator (DEC): 4.6, 7.17
DMR: 3.5
Dry adiabatic lapse: 5.1

E

EchoLink: 3.4
Emergency Coordinator (EC): 1.1, 4.6, 7.17
    Report forms: 7.17
Emergency operations center (EOC): 1.4, 1.6
    Safety: 2.5
Enhanced Fujita Scale: 5.9
Equipment: 3.1ff
Exercises and drills: 4.6

Index   1

F

Facebook: 1.6, 3.12, 3.25, 7.6
False statement notice: 7.8, A5.1
Federal Emergency Management Agency (FEMA)
    Incident Command System (ICS): 4.2
    Integrated Public Alert and Warning System
        (IPAWS): 3.9
    National Incident Management System (NIMS): ... 4.2
    Smartphone app: 3.17
    Training courses: 4.2
FEMA App: 3.17
First aid: 4.3
Fixed location storm spotting: 2.4ff
    Safety: 2.4
Flash flood: 5.5
Flooding: 5.5
Fog: 5.18

G

Go-kit: 3.13, 7.2
    Digital modes: 3.5
Google Earth: 3.8, 3.22, A6.1
GPS receivers: 3.7
Graham, Kenneth, WX4KEG: 6.8
GRLevelX: 3.23

H

Hail: 5.6
Hazardous Weather Outlook: 7.4
History: 1.2ff
    Cooperative Observer Program: 1.3
    NADWARN: 1.3
    National Weather Service (NWS): 1.2ff
    SKYWARN: 1.3
    Storm spotting today: 1.5
    Weather Bureau: 1.3
Hurricane force: 6.3
Hurricane Watch Net (HWN): 6.7
Hurricanes: 6.1ff
    Georges (1998): 6.13
    Gustav (2008): 6.14
    Hortense (1996): 6.12
    In Puerto Rico: 6.11
    Katrina (2005): 6.17
    Lenny (1999): 6.14
    Luis (1995): 6.11
    Maria (2017): 6.23
    Saffir-Simpson Scale: 6.2
    Sandy (2012): 6.18

I

Image sharing
    Apps: 3.12
    Cloud services: 3.12
    WhatsApp: 3.12
Incident Command System (ICS): 4.2, 7.14
Instagram 3.12
Instant messages: 3.9
Integrated Public Alert and Warning System
    (IPAWS): 3.9, 7.10
Integrating mapping and weather data: A6.1
Interactive NWS (iNWS)
    iNWS Alert: 3.17
    iNWS Mobile: 3.17
    iNWS Mobile Web: 3.17
Internet Radio Linking Project (IRLP): 3.4
Internet tools: 7.10

J

JetStream - Online School for Weather: 4.4
Joplin, Missouri tornado: 5.10

L

Landslide: 5.19
Lifting condensation level (LCL): 5.2
Lighting: 3.13
Lightning: 5.4
    Protection: A8.1
Local Storm Report (LSR): 7.12

M

Magazines: 4.5
Maps: 3.8
    Offline Maps & Navigation: 3.8
Memorandum of Understanding between NWS
    and ARRL: A4.1
Mesonet: 7.11
Messaging apps: 3.9
MetEd professional development series: 4.4
Microburst: 5.6
Mobile operation 2.2ff
    Antennas: 3.2
    APRS: 3.4
    Internet capability: 3.9
    Lighting: 3.13
    Navigation aids: 3.7
    Preparations for activation: 7.2
    Radios: 3.1
    Safety: 2.2
Multi-cell thunderstorm: 5.3
MyRadar Weather Radar: 3.22

N

NADWARN: 1.3
National Hurricane Center (NHC): 4.6, 6.6ff
    Advisories: 6.3
    Graham, Kenneth, WX4KEG: 6.8
    WX4NHC amateur station: 6.6, A7.1ff
National Hurricane Conference: 4.4
National Incident Management System (NIMS): 4.2
National Warning System (NAWAS): 7.11
National Weather Service (NWS): 1.1, 3.2
    Advanced Weather Interactive Processing
        (AWIPS): 4.4
    AWARE magazine: 4.4
    Climate Prediction Center: 4.4

Hazardous Weather Outlook: 7.4
JetStream - Online School for Weather: 4.4
Memorandum of Understanding with ARRL: A4.1
MetEd professional development series: 4.4
NWS Chat: 7.4, 7.10
Radar: 5.12
Severe thunderstorm definition: 5.3
Special Weather Statement: 7.3
Storm Prediction Center (SPC): 3.16
Training: 4.3
Weather Safety Information: 2.6
Nets: 2.2, 2.6, 3.1ff, 4.6, 6.6
   Hurricane Watch Net (HWN): 6.7
   SKYWARN: 3.1, 3.4, 3.6, 7.6
   VoIP Hurricane Net: 6.10
NOAA Weather Radio: 3.2ff
   Broadcastify: 3.3
   Frequencies: 3.3
   Outage map: 3.3
   SAME message: 3.3
NOAA Weather Unofficial: 3.23
Nor'easter: 5.17
NWS Chat: 7.4, 7.10

P

Precipitation: 5.16
Public Service Activity Report: 7.16

R

Radar: 5.12
RadarScope: 3.22, 3.23
Radios: 3.1
Radiosonde: 5.2
Reportable criteria: 6.6

S

Safety: 2.1ff
   Emergency operations center (EOC): 2.5
   Fixed location: 2.4
   Mobile operation: 2.2
   NWS Weather Safety Information: 2.7
   Spotter Safety Officer (SSO): 2.8
Saffir-Simpson Hurricane Scale: 6.2
SAME message: 3.3
Satellites: 5.13
Section Emergency Coordinator (SEC): 4.6, 7.17
Section Manager (SM): 4.6
Severe weather reports: 7.9
Simulated Emergency Test (SET): 4.6
Single cell thunderstorm: 5.3
SKYWARN: 1.1, 1.5ff, 4.6, 7.3, 7.11, 7.13ff
   Activation: 7.6
   Activation process: 7.5
   D-STAR Nets: 3.6
   Facebook pages: 7.6
   History: 1.3
   Nets: 3.1, 3.4
   Operations Manual: A3.1ff
   Reportable criteria: 6.6
   Social media: 7.6
   Training: 1.6, 4.1ff
   Volunteers: 7.14
   Websites: 3.14
SKYWARN Recognition Day: 1.5
Sleet: 5.17
Smartphones
   Camera: 3.11
   Messaging apps: 3.9
Snowfall: 5.16
Social media: 1.6, 3.12, 3.25, 6.10, 7.5ff, 7.10
Special Weather Statement: 7.3
Spotter Network, The: 3.21
Spotter Safety Officer (SSO): 2.8
Squall line: 5.3
Storm chaser: 1.8
Storm Data (SD) reports: 7.13
Storm Events database: 7.13
Storm Prediction Center (SPC): 3.16
Storm Spotter: 1.6ff
   20 questions: 1.6
   Safety: 2.1ff
Storm spotter
   Activation: 7.1ff
   Antennas: 3.2
   Preparation: 7.1
   Radios: 3.1
   Reportable criteria: 6.6
   Role during activation: 7.7ff
   vs storm chaser: 1.7
Stormpulse: 3.19
Straight-line winds: 5.6
Stress management: 7.15
Supercell thunderstorm: 5.4

T

Thunderstorms
   Multi-cell: 5.3
   Single cell: 5.3
   Supercell: 5.4
Today Weather: 3.23
Tornadoes: 5.10ff
   Enhanced Fujita Scale: 5.9
   Joplin, Missouri: 5.10
   Vortex signature: 5.13
   Waterspout: 5.10
Training: 4.1ff, 7.15
   Exercises and drills: 4.6
Tropical cyclone: 6.4
Tropical depression: 6.3
Tropical disturbance: 6.3
Tsunami: 5.20
Twitter 1.6, 3.12, 3.25, 6.10, 7.6, 7.10

V

Videos: 4.5
VoIP Hurricane Net: 6.10
Volcanic activity: 5.19
Volunteers: 7.14

W

War Emergency Radio Service (WERS): 1.4
Warning Coordination Meteorologist (WCM): 5.15
Waterspout: 5.10
Weather Bureau: 1.3
Weather Channel, The: 3.20
   Apps: 3.23
Weather forecast: 7.1
Weather Message: 3.25
Weather Nation: 3.20
Weather radar: 5.12
Weather satellites: 5.13
Weather Spotter's Field Guide (NWS): 4.1
Weather stations: 3.26
Weather Underground: 3.20
   Apps: 3.23
Weather websites: A2.1
Weather World 2010 (WW2010) project: 4.4
WeatherBug: 3.18, 3.23
WeatherTAP: 3.21
Webcam: 7.12
Websites: 3.18ff
WhatsApp: 3.12
Wildfire: 5.19
Wind shear: 5.7
Windy.com: 3.23
Wireless Emergency Alert (WEA): 3.9
WX1BOX: 1.6, 6.10, 7.6, 7.16
WX4NHC: 3.5, 6.6, 6.10, A7.1

Y

Yaesu System Fusion: 3.5